Latest Research on Energy Recovery

Edited by Petrica Vizureanu

Published in London, United Kingdom

Latest Research on Energy Recovery
http://dx.doi.org/10.5772/intechopen.97950
Edited by Petrica Vizureanu

Contributors
Jung-In Yoon, Chang-Hyo Son, Sung-Hoon Seol, Ji-Hoon Yoon, Hsini Mohamed, Souhir Bouzidi, Petrica Vizureanu, Madalina-Simona Baltatu, Andrei-Victor Sandu, Dragos Cristian Achitei, Dumitru-Doru Burduhos-Nergis, Manuela-Cristina Perju, Nguyen Van Toan, Truong Thi Kim Tuoi, Nguyen Huu Trung, Khairul Fadzli Samat, Nguyen Van Hieu, Takahito Ono, Mohsen Sarafraz, Bernard Rolfe, Farid Christo, Ali Mohammadi, Anthony Anukam, Ali Soofastaei, Milad Fouladgar

First published in London, United Kingdom, 2023 by IntechOpen
IntechOpen is the global imprint of INTECHOPEN LIMITED, registered in England and Wales, registration number: 11086078, 5 Princes Gate Court, London, SW7 2QJ, United Kingdom

British Library Cataloguing-in-Publication Data
A catalogue record for this book is available from the British Library

Additional hard and PDF copies can be obtained from orders@intechopen.com

Latest Research on Energy Recovery
Edited by Petrica Vizureanu
p. cm.
Print ISBN 978-1-80355-609-3
Online ISBN 978-1-80355-610-9
eBook (PDF) ISBN 978-1-80355-611-6

Meet the editor

Petrica Vizureanu obtained an MSc and Ph.D. in Heating Equipment at the "Gheorghe Asachi" Technical University, Iasi, Romania in 1992 and 1999, respectively. Dr. Vizureanu is currently a full professor and scientific supervisor in Materials Engineering at the same university and the appointed director of the Department of Technologies and Equipment for Materials Processing. He is an editor and guest editor for many journals and publishing houses. His research focuses on expert systems for heating system programming, computer-assisted design for heating equipment, heating equipment for materials processing, heat transfer, biomaterials, and geopolymers. He has published more than 200 papers in international journals and conference proceedings and 40 books.

Contents

Preface

While great efforts are being made to reduce energy consumption, the world needs to further explore renewables like solar, wind, hydropower, geothermal, and even sustainable nuclear energy. However, due to high capital cost, these alternatives have not found wide application in industry. However, this is beginning to change with the emergence of scientific evidence that alternative sources of energy can help achieve sustainability in energy recovery (ER) processes. The case for other industries will follow the same argument especially since most of these heavy industries require enormous amounts of energy, derivable from fossil fuels oil and gas due to their resilience in offering likewise enormous motive power twenty-four hours a day. This book discusses sustainable ER development in our deliberate pursuit of energy independence in a world bedeviled by the ill effects of environmental degradation. In a world that needs energy, we should embrace the practice of utilizing ER systems in all their forms, natural and technological.

This book investigates ER in large industries and utilities throughout the world. In all aspects of heat exchanges, energy recoveries are involved, but the problem in all energy recoveries is that the percentage of energy recovered is not related to the total energy consumption in industry, organization, or household. It should be noted that among other objectives of studying ER one should take cognizance of the fact the recovered energy has these attributes: (1) it is quality energy because it is immediately useful without further processing depending on the point of collection, either through recuperators or economizers, whereby utilization is in preheating incoming air to a combustion chamber/ furnaces or in an AC/refrigeration system, which depends on the outside; (2) it becomes available energy, serving the purpose of extracting tangible heat (enthalpy) from the air, which is a two-component system for immediate use in effecting cooling on the area. The explanation here is that the wet air that contains water vapor after reaching its saturation vapor pressure undergoes the process of evaporation thereby cooling that point.

Chapter 1 pays special attention to the magnetocaloric effect (MCE) that is observed when magnetic systems are subjected to an external magnetic field. The authors use the derivative of Gibbs free energy to estimate the magnetic entropy change and the mean-field theory to sort out the spontaneous magnetization from the dependence of magnetic entropy change on magnetization vs. Gd systems. The obtained spontaneous magnetization values are in good agreement with those found from the extrapolation from the Arrott plots (H/M vs. M2). An excellent agreement has been found between the values estimated by Landau's theory and those obtained using the classical Maxwell relation. The intermetallic Gd_3Ni_2 and Gd_3CoNi exhibit a large MCE manifested with a high entropic peak that can boost refrigerant capacity, which makes these alloys good candidates for magnetic refrigerators.

A micro-thermoelectric generator (TEG) possesses great potential for powering wireless Internet of Things (IoT) sensing systems due to its capability of harvesting thermal energy into usable electricity. Chapter 2 reviews the progress in recent studies on micro-TEG from material synthesis to device fabrication. Thermoelectric materials are synthesized by the electrochemical deposition method. Three kinds of high-performance thermoelectric materials, including thick bulk-like thermoelectric material, Pt nanoparticles embedded in a thermoelectric material, and Ni-doped thermoelectric material, are presented.

The growth of the semiconductor market and advancement of manufacturing technology have led to an increase in wafer size and highly integrated semiconductor devices. The temperature of the supplied cooling medium from the chiller that removes the heat produced in the semiconductor manufacturing process is required to be at a lower level because of the high integration. The Joule-Thomson cooling cycle, which uses a Mixed Refrigerant (MR) to produce the cooling medium at a level of –100°C required for the semiconductor process, has recently gained attention. Chapter 3 discusses this in detail. When an MR is used, the chiller's performance is heavily influenced by the composition and proportions of the refrigerant charged to the chiller system. Therefore, this chapter introduces a cooling cycle that uses an MR to achieve the required low temperature of –100°C in the semiconductor manufacturing process and provides the results.

Chapter 4 explains that the solar furnace supply doesn't make changes inside a material by using electric energy produced by a photovoltaic system that converts solar energy. The performances of a solar furnace used in various applications from industry are influenced by various factors. One of these factors imposes the acquisition of certain large densities of the radiant power and requires a geometric form of the concentrator. The most important research is made on the behavior of some metallic alloys at elevated temperatures, on some purifying materials, and on the achievement of some chemical synthesis. For manufacturing electrothermal furnaces, a series of specific materials are used that are necessary for the achievement of the furnace chamber (e.g., heating elements and measurement systems of the temperature).

Since the world is gradually drifting toward sustainable development, renewable energy technologies are gaining traction and gasification technology is one of many renewable energy technologies that have gained popularity in recent times. Gasification technology is one of three main (combustion and pyrolysis) thermo-chemical conversion pathways that can be used to recover energy from biomass materials. Chapter 5 presents an overview of gasification technology and discusses the different types of gasification systems that are commonly used today for the recovery of energy. The limitations of each type of gasifier in relation to performance and feedstock conversion are also discussed, including research priority areas that will allow for system optimization in terms of efficiency.

Chapter 6 proposes a novel, clean thermochemical process that harnesses thermal plasma technology to co-produce hydrogen and ammonia using a chemical looping process. The thermodynamic potential and feasibility of the process are demonstrated using a simulation of the system with aluminium and aluminium oxide as the oxygen and nitrogen carriers between the reactors. The effect of different operating parameters, such as feed ratio and temperature of the reactor, on the energetic performance of the process is investigated. It is demonstrated that the process can operate at an approximate self-sustaining factor of 0.11 and an exergy partitioning fraction of up to 0.65. Integrating the process with solar photovoltaics shows a solar share of 32% without considering any battery storage units.

Finally, Chapter 7 provides an overview of energy efficiency in the mining industry with a particular focus on the role of fuel consumption in hauling operations in mining. Moreover, as the costliest aspect of surface mining with a significant environmental impact, diesel consumption is investigated in this chapter. This research seeks to develop an advanced data analytics model to estimate the energy efficiency of haul trucks used in surface mines, with the goal of lowering operating costs. The visualized results also clarify the general minimum areas in the plotted fuel consumption graphs. These areas potentially open a new window for researchers to develop optimization models to minimize haul truck fuel consumption in surface mines.

Petrica Vizureanu
"Gheorghe Asachi" Technical University,
Iasi, Romania

Section 1

Heat Recovery

Chapter 1

Magnetocaloric Properties in Gd_3Ni_2 and Gd_3CoNi Systems

Mohamed Hsini and Souhir Bouzidi

Abstract

Intermetallic Gd_3Ni_2 and Gd_3CoNi undergo second-order ferromagnetic paramagnetic phase transition at the Curie temperature, T_C. They exhibit a large magnetocaloric effect (MCE). This MCE is manifested with a high entropic peak of 8 and 8.3 $J.Kg^{-1}\ K^{-1}$, at the vicinity T_C under 5 T magnetic applied field for Gd_3Ni_2 and Gd_3CoNi, respectively. With their boosted MCE and large refrigerant capacity, Gd_3Ni_2 and Gd_3CoNi compounds can be a candidate as a magnetocaloric refrigerator which is still one of the current research projects recommended by the low energy consumption and low environmental impact of these devices. Based on the Landau theory, Gibb's free energy leads to determine temperature-dependent parameters which correspond to the electron condensation energy and magnetoelastic coupling and the magnetic entropy change which is a very crucial parameter to evaluate the MCE of a given magnetic system.

Keywords: magnetic energy, magnetic entropy change, magnetization, phase transition

1. Introduction

The study of magnetic materials having boosted magnetocaloric effect (MCE) and large refrigerant capacity applied in low- and room-temperature magnetocaloric refrigerators is one of the current research projects recommended by the low energy consumption and the safe environmental impact of these materials [1–5]. The MCE is observed when magnetic systems are subjected to an external magnetic field. For a ferromagnet, in an adiabatic process, the MCE presents itself as follows: when an external magnetic field is applied to the ferromagnet the temperature increases and decreases when this magnetic field is removed. From this, a famous quantity can characterize the MCE which is the magnetic entropy change, $-\Delta S_M$. From$-\Delta S_M$, one may evaluate the refrigerant capacity of the material, which expresses the exchanged heat in a thermodynamic cycle of magnetic refrigerators. The optimization and development of magnetic refrigerator devices depend on a solid thermodynamic description of the magnetic material, and its properties throughout the steps of the cooling cycles [6, 7]. Among these magnetic systems, intermetallic alloys formed by rare earth (R) and transition metal (M) such as Gd_3Ni_2, Lu_2Pd_5 and $Nd_2Co_{1.7}$ [8–10]. The projected applications of these materials include magnetic refrigeration, magnetic memory, spintronics and magnetic sensors, etc. Rare earth intermetallic alloys have been attractive for researchers because of their richness in their role in a variety of applications and fundamental physics. Due to their highly localized unpaired 4f electrons, the rare-earth ions in these solid systems can retain their atomic moments. As a result, a very large magnetic moments range can be established. Also, the rare-earth atoms in these samples are heavy, the spin-orbit interaction dominates to be responsible for strong magneto crystalline anisotropy [11]. Recently, Provino et al. [12] reported the crystal structure, thermal stability, magnetic behavior and MCE of Gd_3Ni_2 and Gd_3CoNi compounds.

Generally, SO ferromagnetic-paramagnetic (FM-PM) phase transitions are one of the vital issues related to the functionalities and fundamental physics of magnetic systems. The Landau theory for phase transitions was used to describe the MCE in Gd_3Ni_2 and Gd_3CoNi systems with magnetoelastic and magnetoelectronic couplings [13–16]. As shown in the work of Provino et al. [12], for Gd_3Ni_2 and Gd_3CoNi compounds, the applied magnetic field, H, dependence on peaks of $-\Delta S_M$ obtained near the Curie temperature, T_C, increases proportionally ($-\Delta S_M \sim H^{\frac{2}{3}}$). This behavior is matching with the mean-field theory (MFT). Moreover, the MFT establishes relations between $-\Delta S_M$ and magnetization, M [17]. In addition, the theory of critical phenomena indicates the existence of universal magnetocaloric behavior in materials undergoing SO FM-PM [18–20]. However, the critical exponents can set the behavior magnetic phase transitions.

Since Gd_3Ni_2 and Gd_3CoNi magnetic materials can be described by the MFT, we chose to study, in this paper, the MCE of these samples using both the Landau model and MFT. These two approaches provide side by side the estimation of both spontaneous magnetization, M_s and $-\Delta S_M$. First, we used the $-\Delta S_M$ values deduced from isothermal magnetization measurements to sort out the M_s using the MFT. Results are then compared with those determined from the Arrott plots extrapolation ($\frac{H}{M}$ vs. M^2). Second, the Landau theory was applied to estimate the Gibbs free energy, G and $-\Delta S_M$ near T_C. Generated $-\Delta S_M$ results were compared with the ones estimated using the classical Maxwell relation.

2. Theory

Based on the Landau theory, the Gibb's free energy reads as [15]:

$$G(T,M) = G_0 + \frac{1}{2}A(T)M^2 + \frac{1}{4}B(T)M^4 + \frac{1}{6}C(T)M^6 - MH \tag{1}$$

where the coefficients $A(T)$, $B(T)$ and $C(T)$ are temperature-dependent parameters that correspond to the electron condensation energy and magnetoelastic coupling, M is the magnetization and H is the magnetic applied field. At the equilibrium condition, $\frac{\partial G}{\partial M} = 0$, the magnetic equation of the state is obtained as:

$$\frac{H}{M} = A(T) + B(T)M^2 + C(T)M^4 \tag{2}$$

The magnetic entropy is obtained as:

$$-\Delta S_M(T,M) = \left(\frac{\partial G(H,T)}{\partial T}\right)_H = \frac{1}{2}A'M^2 + \frac{1}{4}B'M^4 + \frac{1}{6}C'M^6 \tag{3}$$

where $A' = \frac{\partial A}{\partial T}$, $B' = \frac{\partial B}{\partial T}$ and $C' = \frac{\partial C}{\partial T}$.

According to the renormalization group approach to scaling, Dong et al. [21] have reported that the zero-field spontaneous magnetization, M_s, Consequently, $-\Delta S_M$ should not be null. Then, Eq. (3) can be rewritten as:

$$-\Delta S_M(T,M) = \frac{1}{2}A'\left(M^2 - M_s{}^2\right) + \frac{1}{4}B'\left(M^4 - M_s{}^4\right) + \frac{1}{6}C'\left(M^6 - M_s{}^6\right) \qquad (4)$$

To estimate the zero-field spontaneous magnetization, M_s, we have a look on the expression of the magnetic entropy from the mean-field theory [9]:

$$S(\sigma) = -Nk_B\left[ln\,(2J+1) - ln\left[\frac{sinh\left(\frac{2J+1}{2J}B_J^{-1}(\sigma)\right)}{sinh\left(\frac{1}{2J}B_J^{-1}(\sigma)\right)}\right] + B_J^{-1}(\sigma)\sigma\right] \qquad (5)$$

where N is the number of magnetic moments, k_B is the Boltzmann constant, J is the angular spin value, σ is the reduced magnetization ($\sigma = \frac{M}{M_0}$, $M_0 = NJg\mu_B$: saturation magnetization) and B_J is the Brillouin function for a given J value. For small M values, Eq. (5) can be performed using a power expansion, and the magnetic entropy change$-\Delta S_M$ is proportional to $\sigma^2\left(M^2\right)$:

$$-\Delta S_M(\sigma) = \frac{3J}{2J+1}Nk_B\sigma^2\left(M^2\right) + O\left(\sigma^4\right) \qquad (6)$$

Below T_C, the ferromagnetic materials acquire M_{spont}, as a result, the $\sigma = 0$ state is never reached. Then, the contribution of the reduced spontaneous magnetization $\sigma_{spont} = \frac{M_{spont}}{M_0}$ should be added. Consequently, if we consider only the first term of Eq. (2), the magnetic entropy change may be written as:

$$-\Delta S_M(\sigma) = \frac{3}{2}\frac{J}{(J+1)M_0}Nk_B\left(M^2 - M_s{}^2\right) \qquad (7)$$

3. Results and discussions

Figure 1 presents the isothermal $-\Delta S_M$ vs. M^2 plots, in the ferromagnetic region ($T < T_C$). Curves (black symbols) present horizontal drift from the origin, corresponding to the value of $M_s{}^2(T)$.

As shown in **Figures 1** and **2**, all curves at different temperatures obey the same regularity and a series of linear dependence with an approximately constant slope occurs. This indicates that it is possible to analyze the current experimental results with the mean-field theory.

Linear fits are applied on the isothermal -ΔS_M vs. M^2 plots inside the ferromagnetic region to sort out M_s. The same stuff is following to obtain M_s from the Arrott plots: $\frac{H}{M}$ vs. M^2 in **Figure 2** for the Gd_3Ni_2 and Gd_3CoNi systems.

Figure 3 shows practically the same curve M_s vs. T from $-\Delta S_M$ vs. M^2 (red symbols) and from Arrott plots (black symbols).

As seen in **Figure 3**, as the temperature decreases, the spontaneous magnetization becomes larger, suggesting that the systems are approaching a spin ordering state and a strong localization of moments is formed.

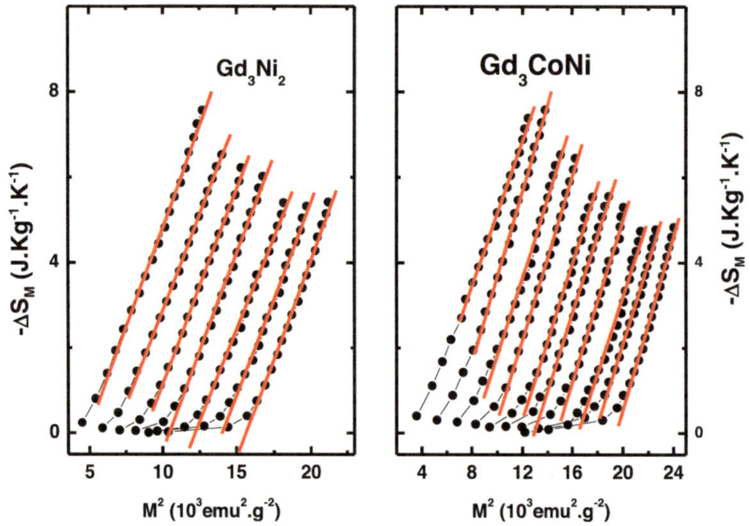

Figure 1.
Linear fits of $-\Delta S_M$ vs. M^2 *plots for the* Gd_3Ni_2 *and* Gd_3CoNi *alloys.*

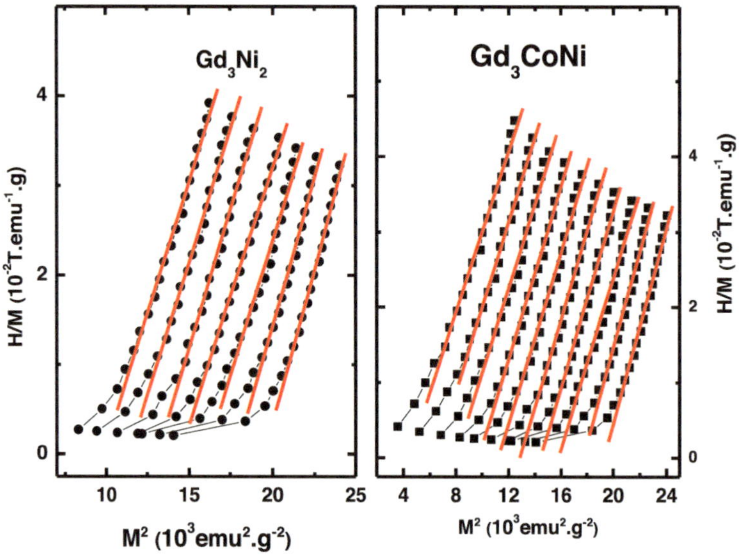

Figure 2.
Linear fits of $\frac{H}{M}$ *vs.* M^2 *plots for the* Gd_3Ni_2 *and* Gd_3CoNi *alloys.*

Based on the scaling hypothesis, the critical exponent, β, the reduced temperature, $\varepsilon = \frac{T-T_C}{T_C}$, and the saturation magnetization, M_0, as [22]:

$$M_{spont}(T) \propto M_0(-\varepsilon)^{\beta}. \tag{8}$$

By changing Eq. (8) to log–log scale, the value of β corresponds to the slope of the curve $\ln(M_s)$ vs. $\ln(-\varepsilon)$ in **Figure 4**.

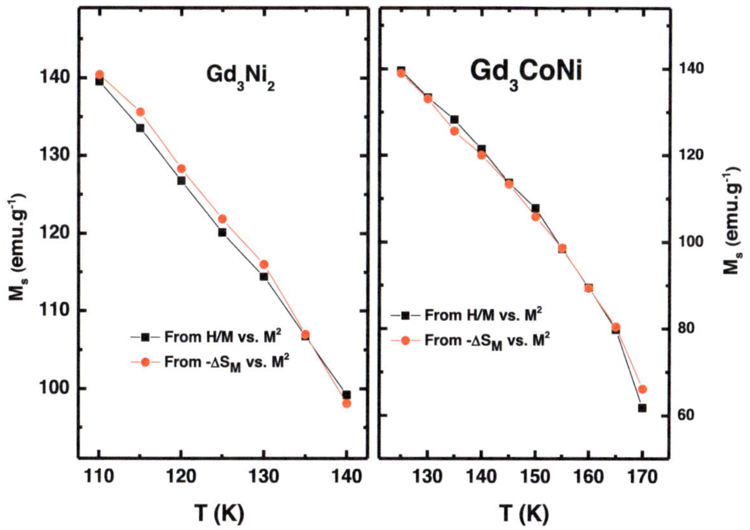

Figure 3.
M_s vs. T from $-\Delta S_M$ vs. M^2 (red symbols) and from Arrott plots (black symbols) for the Gd_3Ni_2 and Gd_3CoNi samples.

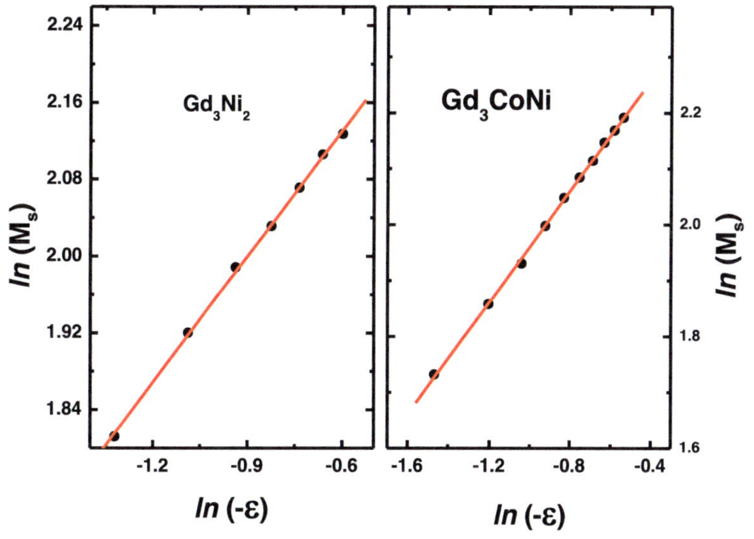

Figure 4.
Linear fit of $ln\,(M_s)$ vs. $ln\,(-\varepsilon)$ for the Gd_3Ni_2 and Gd_3CoNi samples.

The value of the exponent β, is found to be 0.49 with Gd_3Ni_2 and 0.47 with Gd_3CoNi. The β values are consistent with the standard mean-field model (β = 0.5 [22]).

In the next, Fitting the Arrott plots in **Figure 5** gives the parameters $A(T)$, $B(T)$, shown in **Figure 6**, and $C(T)$ shown in **Figure 7** for the Gd_3Ni_2 and Gd_3CoNi compounds.

The $A(T)$ curve is positive and would get a minimum value at the vicinity of T_C. On the other hand, the magnetic phase transition order is governed by the sign of $B(T)$

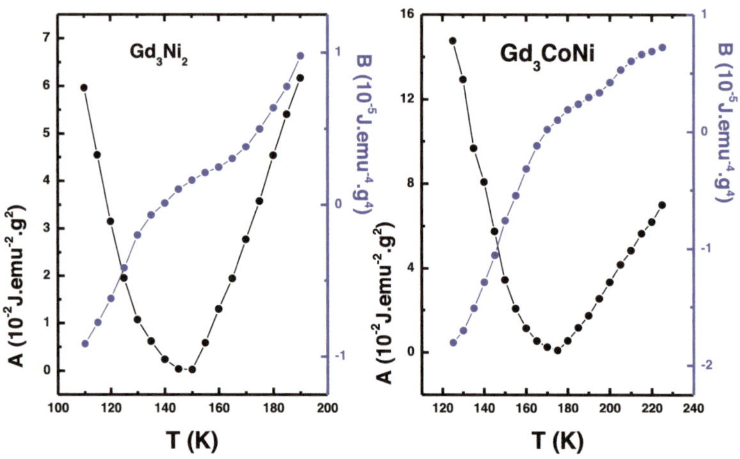

Figure 5.
Temperature dependence of Landau coefficients A, B for the Gd_3Ni_2 *and* Gd_3CoNi *samples.*

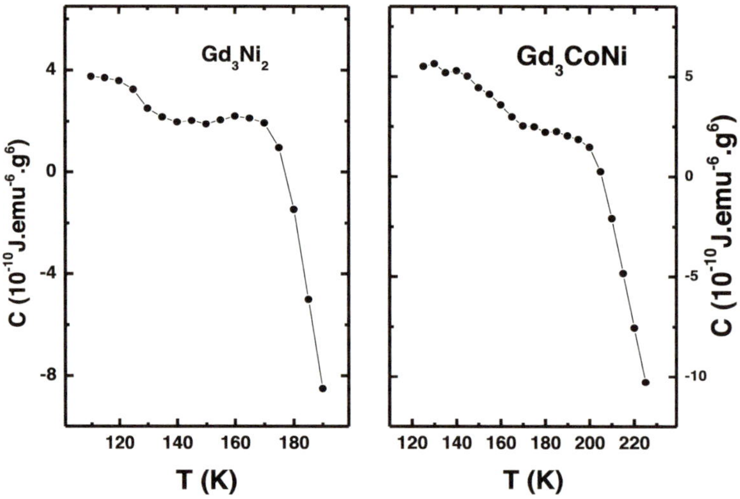

Figure 6.
Temperature dependence of Landau coefficient C for the Gd_3Ni_2 *and* Gd_3CoNi *samples.*

at the transition: a SO occurs when $B(T) \geq 0$ while a first-order transition happens if $B(T) < 0$. In this work, the positive sign of $B(T_C)$ indicates a SO magnetic phase transition for the Gd_3Ni_2 and Gd_3CoNi compounds. Besides, *C(T)* is positive at T_C but in other cases, it is negative or positive. After sorting $A(T)$, $B(T)$, and $C(T)$, Gibb's free energy change, $\Delta G = G - G_0$ can be estimated. The temperature dependence of ΔG under $H = 1$ to 5 T is plotted in **Figure** 7.

As shown in **Figure** 7, $\Delta G(T)$ changes quickly from the high absolute values to the low ones while going from the FM to PM region. The first principle of thermodynamics indicates that there is conservation of energy and in this case, if the internal energy of the system varies, it is because there is an exchange of energy with the external environment either in the form of work or in the form of heat.

The temperature dependence of $(-\Delta S_M)$ is calculated using Eq. (4), under magnetic field varying from 1 to 5 T.

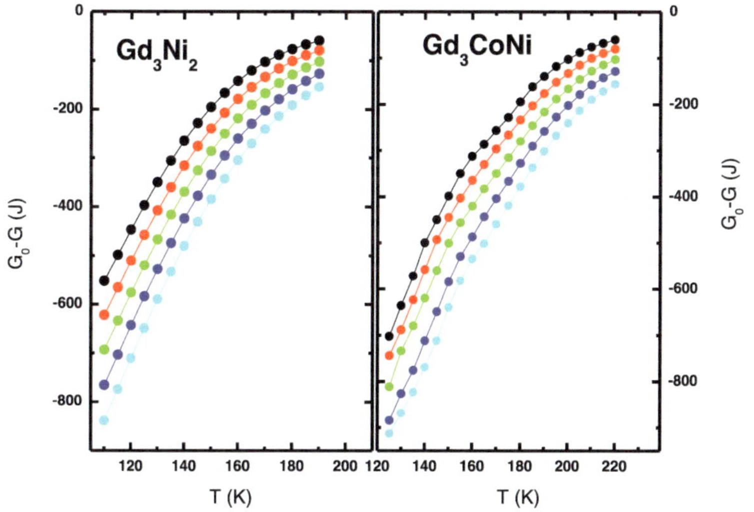

Figure 7.
Temperature dependence of $\Delta G = G - G_0$ under $H = 1$ to 5 T for the Gd_3Ni_2 and Gd_3CoNi alloys.

Figure 8 shows a good agreement between the Landau plots (red lines) and the experimental plots of $(-\Delta S_M)$ vs. T for the Gd_3Ni_2 and Gd_3CoNi alloys.

For the Gd_3Ni_2 and Gd_3CoNi compounds, the peak of $-\Delta S_M$ is achieved near their T_C. Under 1 T applied magnetic field, the entropic peak is about 2.1 and 2.3 $J.Kg^{-1}\ K^{-1}$ or under 5 T, it increases to be 8 and 8.3 $J.Kg^{-1}\ K^{-1}$ for Gd_3Ni_2 and Gd_3CoNi, respectively. These alloys exhibit relatively large MCE at intermediate temperatures. The $-\Delta S_M$ is not the only parameter to quantify the potential of a magnetic refrigerant: the cooling power or the refrigerant capacity, RC, is another important quantity. The RC quantifies the efficiency of a magnetic system in terms of the energy transfer between the cold and the hot reservoir in a perfect thermodynamic refrigeration cycle.

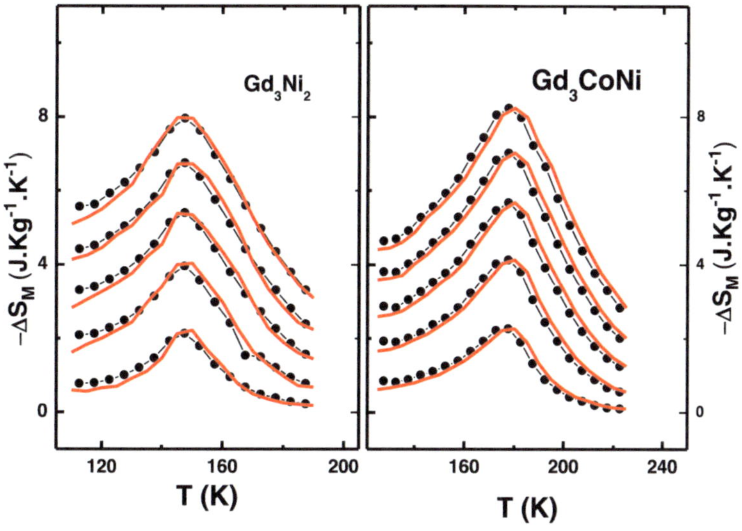

Figure 8.
Comparison between experimental (black symbols) and simulated (red lines) $-\Delta S_M(T)$ using Landau theory under $H = 1$ to 5 T for the Gd_3Ni_2 and Gd_3CoNi compounds.

It estimates the transferred heat between the hot and the cold ends; so, for practical applications, a boost RC over a wide temperature range coupled with hight MCE is desirable. The RC values can be calculated by integrating the area of the $-\Delta S_M$ vs.T curves (between T_1 and T_2) as $RC = \int_{T_1(cold)}^{T_2(hot)} |\Delta S_M| dT$, where T_1 and T_2 represent the temperatures of the hot and cold reservoir, respectively. This returns to assume T_1 and T_2 the low and hot temperatures, respectively, at full width at half maximum (FWHM) of $-\Delta S_M$. For the two compounds, the RC values are of the order of about 540 JKg^{-1} under 5 T magnetic field. This high RC selects the intermetallic Gd_3Ni_2 and Gd_3CoNi compounds as a good magnetic refrigerator.

4. Conclusion

In this work, we used the derivative of the Gibbs free energy to estimate the magnetic entropy change and the mean-field theory to sort out the spontaneous magnetization from the dependence of magnetic entropy change on magnetization, $(-\Delta S_M)$ vs. M^2 in Gd_3Ni_2 and Gd_3CoNi systems. The obtained spontaneous magnetization values are in good agreement with those found from the extrapolation from the Arrott plots (H/M vs. M^2). An excellent agreement has been found between the $(-\Delta S_M)$ values estimated by Landau theory and those obtained using the classical Maxwell relation. The intermetallic Gd_3Ni_2 and Gd_3CoNi exhibit a large MCE manifested with hight entropic peak at the vicinity of T_C and boost refrigerant capacity which make these alloys as a good candidate for magnetic refrigerator.

Conflict of interest

The authors declare no conflict of interest.

Author details

Mohamed Hsini* and Souhir Bouzidi
University of Monastir, Monastir, Tunisia

*Address all correspondence to: mohamed.hsini.14@gmail.com

References

[1] Feng J, Li F, Wang G, Wang JQ, Hu J. Magnetocaloric effect in ercu-based metallic glass composite. Journal of Non-Crystalline Solids. 2020;**536**:120004. DOI: 10.1016/j.jnoncrysol.2020.120004

[2] da Silva-Santos JA, Santos JM, Plaza EJR, Moreno NO. Magnetocaloric effect in the Nd2TIn8 (T=Rh, Ir) intermetallics. Journal of Alloys and Compounds. 2020;**829**:154413. DOI: 10.1016/j.jallcom.2020.154413

[3] Freitas VLO, Costa SS, Pacheco CJ, Alho BP, Ribeiro PO, von Ranke PJ, et al. Magnetism and magnetocaloric effect in amorphous ferrimagnetic systems: Application to the $Gd_{55}Fe_{45-x}Al_x$ series. Journal of Non-Crystalline Solids. 2021;**571**:121133. DOI: 10.1016/j.jnoncrysol.2021.121133

[4] Bhowmik TK. Observation of the magnetic criticality and the magnetocaloric effect in $LaMn_2Si_2$ in the vicinity of the phase transition around the room temperature. Physics Letters A. 2021;**17**:127724. DOI: 10.1016/j.physleta.2021.127724

[5] Advanced Energy Materials. 2020;**10**:1903741. DOI: 10.1002/aenm.201903741

[6] Tran HB, Fukushima T, Momida H, Sato K, Makino Y, Oguchi T. Theoretical prediction of large anisotropic magnetocaloric effect in MnP. Computational Materials Science. 2021;**188**:110227. DOI: 10.1016/j.commatsci.2020.110227

[7] Loudaini A, Aggour M, Bahmad L, Mounkachi O. Magnetic properties, magnetocaloric effect and cooling performance of $AlFe_2B_2$ compound: Ab initio, Monte Carlo and numerical modeling study. Materials Science and Engineering B. 2021;**264**:114935. DOI: 10.1016/j.mseb.2020.114935

[8] Herrero A, Oleaga A, Manfrinetti P, Provino A, Salazar A. Study of the magnetocaloric effect in intermetallics RTX (R = Nd, Gd; T = Sc, Ti; X = Si, Ge). Intermetallics. 2019;**110**:106495. DOI: 10.1016/j.intermet.2019.106495

[9] Herrero A, Oleaga A, Aseguinolaza IR, Garcia-Adeva AJ, Apiñaniz E, Garshev AV, et al. Tailoring the magnetocaloric, magnetic and thermal properties of Dy6(Fe,Mn)X2 intermetallics (Xdouble bondSb, Te, Bi). Journal of Alloys and Compounds. 2022;**890**:161849. DOI: 10.1016/j.jallcom.2021.161849

[10] Villars P, Cenzual K. Pearson's Crystal Data – Crystal Sructure Database for Inorganic Compounds. OH: ASM International, Materials Park; 2010/11

[11] Gupta S, Lukoyanov AV, Knyazevc YV, Kuz'minc YI, Suresh KG. Field induced metamagnetism and large magnetic entropy change in RRhSi (R = Tb, Dy, Ho) rare earth intermetallics. Journal of Alloys and Compounds. 2021;**888**:161493. DOI: 10.1016/j.jallcom.2021.161493

[12] Provino P, Smetana V, Paudyal D, Gschneidner AK Jr, Mudring AV, Pecharsky VK, et al. Gd3Ni2 and Gd3CoxNi2−x: magnetism and unexpected Co/Ni crystallographic ordering. Journal of Materials Chemistry C. 2006;**4**:6078. DOI: 10.1039/C6TC01035K

[13] Kachniarz M, Salach J. Characterization of magnetoelastic properties of Ni–Zn ferrite in wide range of magnetizing fields for stress sensing

applications. Measurement. 2021;**15**: 108301. DOI: 10.1016/j. measurement.2020.108301

[14] Amaral VS, Amaral JS. Magnetoelastic coupling influence on the magnetocaloric effect in ferromagnetic materials. Journal of Magnetism and Magnetic Materials. 2004;**272–276**: 2104-2105. DOI: 10.1016/j. jmmm.2003.12.870

[15] Ennassiri N, Tahiri N, El Bounagui O, Ez-Zahraouy H, Benyoussef A. Magnetic, magnetocaloric and transport properties in AlCMn3 antiperovskite compound. Journal of Alloys and Compounds. 2018;**741**:1196

[16] Chatterjee A, Majumdar S, Chatterjee S, Dippel AC, Gutowski O, Zimmermann MV, et al. Magnetoelastic coupling at spin-glass-like transition in Sr3NiSb2O9. Journal of Alloys and Compounds. 2019;**77**:30. DOI: 10.1016/j. jallcom.2018.11.074

[17] Amaral JS, Silva NJO, Amaral VS. Estimating spontaneous magnetization from a mean field analysis of the magnetic entropy change. Journal of Magnetism and Magnetic Materials. 2010;**322**:1569-1571. DOI: 10.1016/j. jmmm.2009.09.024

[18] Pramanik AK, Banerjee A. Critical behavior at paramagnetic to ferromagnetic phase transition in Pr0.5Sr0.5MnO3. A bulk magnetization study. Physical Review B. 2009;**79**: 214426. DOI: 10.1103/PhysRevB. 79.214426

[19] Franco V, Blazquez JS, Conde A. Field dependence of the magnetocaloric effect in materials with a second order phase transition: A master curve for the magnetic entropy change. 2006;**89**: 222512. DOI: 10.1063/1.2399361

[20] Franco V, Conde A, Romero-Enrique JM, Blazquez JS. Field dependence of the adiabatic temperature change in second order phase transition materials: Application to Gd. Journal of Applied Physics. 2009;**20**:103911. DOI: 10.1063/1.3261843

[21] Dong Q, Zhang H, Sun J, Shen B, Franco V. Magnetocaloric response of Fe75Nb10B15 powders partially amorphized by ball milling. Journal of Applied Physics. 2008;**103**:116101. DOI: 10.1063/1.3155982

[22] Fisher ME. The theory of equilibrium critical phenomena. Reports on Progress in Physics. 1967;**30**:615. DOI: 10.1088/0034-4885/31/1/508

Chapter 2

Micro-Thermoelectric Generators: Material Synthesis, Device Fabrication, and Application Demonstration

Nguyen Van Toan, Truong Thi Kim Tuoi, Nguyen Huu Trung, Khairul Fadzli Samat, Nguyen Van Hieu and Takahito Ono

Abstract

Micro-thermoelectric generator (TEG) possesses a great potential for powering wireless Internet of Things (IoT) sensing systems due to its capability of harvesting thermal energy into usable electricity. Herein, this work reviews the progress in recent studies on the micro-TEG, including material synthesis, device fabrication, and application demonstration. Thermoelectric materials are synthesized by the electrochemical deposition method. Three kinds of high-performance thermoelectric materials, including thick bulk-like thermoelectric material, Pt nanoparticles embedded in a thermoelectric material, and Ni-doped thermoelectric material, are presented. Besides the material synthesis, novel fabrication methods for micro-TEG can also help increase its output power and power density significantly. Two fabrication processes, micro/nano fabrication technology and assembly technology, are investigated to produce high-performance micro-TEG. Moreover, the fabircated micro-TEG as a power source for portable and wearable electronic devices has been demonstrated successfully.

Keywords: thermal-to-electric energy conversion, micro-thermoelectric generator, thermoelectric materials, micro/nano fabrication technology, assembly technology

1. Introduction

The considerable growth of research studies in energy-harvesting technologies, such as solar energy harvesting [1], RF power harvesting [2], thermoelectric-generator-based electrolyte [3], thermoelectric-generator-based solid thermoelectric materials [4], associated with the Internet of Things (IoT) leads to more demands in the development of the high performance of a micro-thermoelectric generator (TEG). Micro-TEG keeps a role as a charger to the rechargeable battery of IoT sensing systems or even replaces the battery if micro-TEG with high performance is employed. The TEG utilizes the Seebeck effect that can convert thermal energy into electricity. The

TEG has many advantages, including small size, without moving parts, free from noise, greenhouse gases, and long-term operation time [5, 6]. A voltage will be generated once a temperature difference across the micro-TEG is provided.

To enhance the performance of the micro-TEG, high-performance thermoelectric materials and increasing the number of thermoelectric elements are vital factors. Regarding thermoelectric materials, until now, several thermoelectric materials have been studied, including organic materials (metalloporphyrin/single-walled carbon nanotube composite films [7], Poly(3,4-ethylenedioxythiophene) polystyrene sulfonate [8], and compositions of conducting polymers and metal nanoparticles [9]) and inorganic materials (nanoporous silicon [10], cobalt triantimonide [11], bismuth telluride and antimony telluride [12], tin selenide [13], electrodeposited bismuth telluride [14]). Among them, thermoelectric-materials-based BiTe are widely investigated because of their high performance for applications at near room temperature. For synthesis of thermoelectric-materials-based BiTe, several methods have been reported, including thermally evaporated method [15], metal organic chemical vapor deposition method [16], and pulsed laser melting method [17]. Electrochemical deposition is one of the preferred ways to enable the deposited film with high-quality morphology and compactness. Moreover, the electrodeposition method is capable of modifying the morphology, composition, and crystal structure of the synthesized film, which would result in the high performance of the deposited materials. Concerning enhancing the integration density, hundreds of thermoelectric elements could be produced on a small footprint by utilizing micro/nano fabrication technologies; however, some issues still remain. For instance, a complex process is required to create the air bridge between two thermoelectric elements. High contact resistance between thermoelectric elements and substrate results in low-performance micro-TEG. The performance of thermoelectric materials is degraded during their fabrication of the micro-TEG. The height of the thermoelectric element is limited by micro/nano fabrication technology. Thus, it makes micro-TEG low performance and against the practical applications.

In this work, we review the recent progress in the micro-TEG, including material synthesis, device fabrication, and application demonstration. Various high-performance thermoelectric materials synthesized by the electrodeposition method, including thick bulk-like thermoelectric material, Pt nanoparticles embedded in a thermoelectric material, and Ni-doped thermoelectric material, are presented. In addition, the fabrication of micro-TEGs based on micro/nano fabrication technology as well as assembly technology is demonstrated. The performance of the fabricated micro-TEG is compared with other related works. Moreover, the fabricated micro-TEG as a power source for a calculator and a twist watch has been investigated.

2. Basic principles of thermoelectric generator

2.1 Properties of thermoelectric material

2.1.1 Seebeck coefficient

The Seebeck coefficient is defined as the harvested voltage from the temperature difference across the thermoelectric materials. Its standard unit is microvolts per kelvin (μV/K). The Seebeck coefficient may exhibit positive or negative signs, which represents p-type or n-type thermoelectric materials, respectively. The p-type thermoelectric material shows an excess of holes, while the n-type thermoelectric material possesses an excess of free electrons. When a temperature difference appears at the ends of the thermoelectric material block, the charge carriers (electrons or holes) move from the hot side to the cold side, causing a thermoelectric voltage. The following equation depicts the Seebeck coefficient S of thermoelectric materials:

$$S = \frac{\Delta V}{\Delta T}, \tag{1}$$

where ΔV is the voltage gradient between the hot and cold sides of the thermoelectric material, and ΔT is the temperature difference between two sides.

One factor affecting the Seebeck coefficient is charge carrier concentration n. The relationship between the charge carrier and the Seebeck coefficient is proven experimentally and theoretically by published works [18–20].

$$S = \frac{8\pi^2 k_B^2 T}{3eh^2} m^* \left(\frac{\pi}{3n}\right)^2, \tag{2}$$

where k_B is Boltzmann constant, T is temperature, e is the electron charge, h is Planck constant, and m^* is effective mass.

2.1.2 Electrical conductivity

Electrical conductivity is an essential electrical property for thermoelectric material to conduct an electrical current. Electrical conductivity and electrical resistivity are the reciprocals of each other. Macroscopically, electrical conductivity is related to the dimensions and resistance of the measured thermoelectric material, which can be calculated by the following equation:

$$\sigma = \frac{L}{RA}, \tag{3}$$

where L is the length of the material, R is the resistance of the material, A is the contact area perpendicular to the current direction.

In principle, the electrical resistivity of a material characterizes the ability of the material to interrupt electricity flow. Therefore, it is strongly related to the flow of electrons and holes in a material. Those two factors influence the value of electrical conductivity, as shown in the following equation,

$$\sigma = e\,(\mu_e n + \mu_h p), \tag{4}$$

where μ_e, n, μ_h, and p symbolize electron mobility, the carrier density of electron, hole mobility, and carrier density of hole, respectively.

2.1.3 Thermal conductivity

The thermal conductivity k of thermoelectric material is dependent on the charge carriers and the phonon's movement. Generally, the total thermal conductivity of metal increases when the electrical conductivity is high due to the directly proportional relation of electrical conductivity with carrier-charge thermal conductivity.

Therefore, the only option to reduce the thermal conductivity is by scrutinizing the value of lattice thermal conductivity [21, 22]. A lower lattice thermal conductivity results in a smaller value of total thermal conductivity. Introducing the nanoparticles in the metal might reduce the lattice thermal conductivity by blocking the excitation stream of lattice vibration, also known as phonons flow. The interrupted phonons flow increases the phonon scattering and elongates the phonon wavelength. Therefore, the time taken for the heat to transfer will be increased. The total thermal conductivity can be expressed by considering those two factors (charge carriers and lattice), as the following equation,

$$k = k_l + k_e, \tag{5}$$

where k_l and k_e are lattice and charge carrier thermal conductivity, respectively.
Equation of lattice thermal conductivity can be referred to the following relationship.

$$k_l = DC_p\rho, \tag{6}$$

where D, Cp, and ρ signify thermal diffusivity, specific heat, and material density, respectively.

Equation of charge-carriers thermal conductivity is estimated by

$$k_e = ne\mu L_f T, \tag{7}$$

where n is carrier concentration, e is the electron charge, μ is carrier mobility, L_f is Lorenz factor (2.44×10^{-8} WΩK^{-2}), and T is temperature.

2.1.4 Figure of merit

The figure of merit ZT is an instrument to evaluate the performance of thermoelectric materials, which encompassed the factor of the Seebeck coefficient S, electrical conductivity σ, thermal conductivity k, and absolute temperature T of the thermoelectric material. The ZT is defined as follows:

$$ZT = \frac{\sigma S^2 T}{k} \tag{8}$$

To obtain high ZT values of thermoelectric materials, high S and large σ are desired; however, there is a trade-off between S and σ, as shown in Eqs. (2) and (4). Therefore, adjusting the coefficient between S and σ is a critical technique to achieve the highest ZT. Lowering thermal conductivity is also an important point to enhance the ZT, which can avoid the thermal shortcut problem and maintain a large temperature difference between the two sides.

2.2 Thermoelectric generator structure

A TEG is a solid device, which is able to convert thermal energy into electricity or vice versa. It consists of n and p-type thermoelectric elements arranged electrically in series and thermally in parallel. A cross-sectional view and titled view of the TEG structure are shown in **Figure 1(a)** and **(b)**, respectively. It mainly consists of n- and p-type thermoelectric elements, a metal bar, and a substrate.

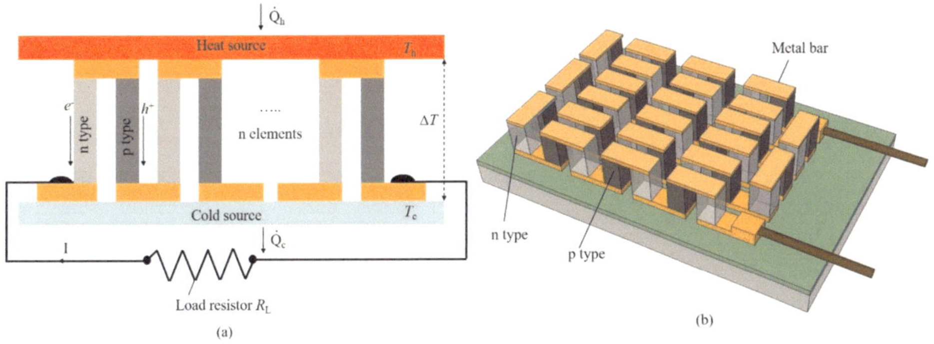

Figure 1.
TEG structure. (a) Cross-sectional view. (b) Titled view.

As mentioned previously, the p-type element has a positive Seebeck coefficient and an excess of holes h^+. The n-type element has a negative Seebeck coefficient and an excess of free electrons e^-. The two elements are connected by an electrical conductor forming a junction, usually a copper strip. When a load resistor R_L is connected in the output terminal of the micro-TEG, an electrical circuit is created. A potential voltage across the resistor is generated once the electrical current flows. The micro-TEG will create the current when a temperature difference across the micro-TEG appears. Higher temperature difference ΔT results in the larger electric output power.

The resistance of the thermoelectric elements is estimated by:

$$R = n\left(\rho_n \frac{L_n}{A_n} + \rho_p \frac{L_p}{A_p}\right) \tag{9}$$

where ρ_n and ρ_p are the electrical resistivity of n and p-type thermoelectric material, respectively, L_n and L_p are the height of n- and p-type thermoelectric elements, respectively, and A_n and A_p are the cross-sectional area of n and p-type thermoelectric elements, respectively.

In the above Eq. (10), the electrical contact resistance is eliminated. However, this resistance is typically quite difficult to be negligible due to the fabrication process. Therefore, the electrical contact resistance R_a should be counted.

$$R = n\left(\rho_n \frac{L_n}{A_n} + \rho_p \frac{L_p}{A_p}\right) + \mathrm{R_a} \tag{10}$$

The generated voltage V_{TEG} could be estimated by the following equation:

$$V_{TEG} = n\left(S_p - S_n\right) \Delta T \tag{11}$$

where n is the number of thermoelectric elements, S_p and S_n are the Seebeck coefficient of p and n types thermoelectric materials, respectively, and ΔT is a temperature difference across the thermoelectric elements.

The maximum electrical output power of the TEG can be calculated by using Eq. (13), which is obtained if a load resistance R_L is equal to the equivalent internal resistance of thermoelectric elements in series [23].

$$P_{max} = \mathrm{n}\frac{\left(S_p - S_n\right)^2 \Delta T^2}{4R_L} = \frac{\mathrm{nA}}{h}\frac{\left(S_p - S_n\right)^2 \Delta T^2}{4\left(\rho_p + \rho_n\right)} \tag{12}$$

where A and h are a cross-sectional surface area and height of thermoelectric elements, respectively. ρ_{p} and ρ_{n} are the electrical resistivities of p-type and n-type thermoelectric materials.

Several factors could affect the performance of the TEGs. Thermoelectric materials with excellent characteristics, including a high Seebeck coefficient, a small electrical resistivity, and a low thermal conductivity, are always desired for enhancing the TEG's performance. Many novel approaches, including utilizing metal nanoparticles [24], nanoporous materials [25], carbon black particles [26], and metal doping [27, 28], have been investigated to improve thermoelectric material's properties. Besides the effects of material properties, selecting proper physical dimensions of thermoelectric elements, such as the width and height of thermoelectric elements, could also contribute to better performance of the TEG [28]. Also, increasing the number of thermoelectric elements would be a valuable method for improving the performance of the TEG, as shown in Eq. (13).

The formula of an electrical energy conversion efficiency η_{TEG} of the TEG [29] is defined by Eq. (14), which indicates that high electrical efficiency of the TEG could be achieved by a high figure of merit ZT as well as a large temperature difference ΔT.

$$\eta_{TEG} = \frac{\Delta T}{T_H}\frac{\sqrt{1+ZT}-1}{\sqrt{1+ZT}+\frac{T_C}{T_H}}, \tag{13}$$

where T_{H} and T_{C} are the hot and cold temperatures of TEG, respectively.

3. Material synthesis

3.1 Electrodeposition method

Thermoelectric materials presented in this work are synthesized by the conventional three-electrode system, which is controlled electrochemically by a potentiostat. The system involves a working electrode, a counter electrode, and a reference electrode. A silicon wafer with Cr-Au layers on the top insulated by SiO_2 layer is employed as a working electrode, while a Pt strip and Ag/AgCl with 3 M KCl solution are utilized as counter and reference electrodes. The synthesized material is formed on the working electrode caused by the oxidation–reduction (redox) reaction. The electrochemical deposition mechanism is quite complicated and has been presented in many publications [30, 31]. It can be summarized as follows. In the electrolyte, the absorbed atom is in the form of the hydrated matter, which is stripped at the interface between the solution and the cathode. Then, it combines with other absorbed atoms to form a new nucleus. This process continues and contributes to the further growth of the deposited material.

One of the benefits of the electrodeposition method is the ability to change the morphology, composition, and crystal structure of deposited film by adjusting certain parameters in the electrodeposition system. All the changes might influence the alteration of the electronic or/and thermal properties of the deposited film. The effectively applied potential on the working electrode is one of the important parameters in the

DOI: http://dx.doi.org/10.5772/intechopen.102649

electrodeposition system that reflect on the variation of the current density. A change of the over potential on the electrode normally affects the current density and a chance to change the morphology.

3.2 Thick bulk-like thermoelectric material

As mentioned in the introduction section, thermoelectric materials could be synthesized by several methods. Although high performance of thin-film thermoelectric materials has been achieved, the TEG produced by thin-film thermoelectric materials possesses a low output power. Once the height of thermoelectric elements is low (a few micrometer heights), it is hard to create a large temperature difference across the TEG device. Thus, its output voltage, as well as output power, is in small value. The evidence could be easily seen via Eqs. (12) and (13). Although an output power of the TEG-utilized thin films could be enhanced by a novel design for heat transfer in a lateral direction, TEG's output power is still not enough for realistic applications. Therefore, a thick film of thermoelectric material with high Seebeck, large electrical conductivity, and low thermal conductivity are always desired to achieve high-performance micro-thermoelectric generators. Typically, thick thermoelectric material films could be formed by a screen printing method, a powder synthesis and sintering method, and a mechanical alloying and spark plasma sintering method; nevertheless, these methods have at least the following disadvantages, such as poor mechanical strength, a high fabrication cost, and low material performance. Herein, we present the thick and stable thermoelectric films synthesized by electrodeposition.

Figure 2 shows the sample preparation process for material synthesis and material evaluation. It starts from a silicon substrate with a thickness of 300 μm (**Figure 2(a)**).

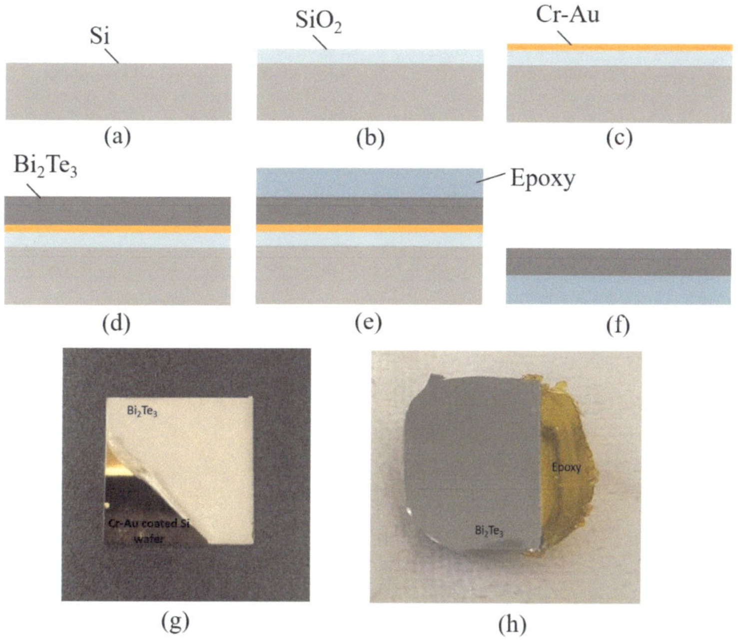

Figure 2.
Sample preparation process. (a) Silicon. (b) SiO_2 deposition. (c) Cr-Au deposition. (d) Thermoelectric material formed by electrodeposition. (d) Epoxy coating. (f) Sample for evaluation.

On top of this substrate, a SiO_2 layer with a thickness of 200 nm is deposited by a plasma-enhanced chemical vapor deposition (PECVD) employing TEOS (TetraEthOxySilan Si $(OC_2H_5)_4$), as shown in **Figure 2(b)**. Next, Cr-Au layers with a thickness of 20 nm and 150 nm are formed on the SiO_2 layer by the sputtering method, respectively (**Figure 2(c)**). The thermoelectric material is subsequently deposited by the electrodeposition method, as discussed in Section 3.1 (**Figure 2(d)**). Because a material property evaluation needs to be conducted on an insulating substrate to avoid short-circuiting, the synthesized films are peeled off from the substrate by epoxy resin, as shown in **Figure 2(e)** and **(f)**. **Figure 2(g)** and **(h)** show the electrodeposited thermoelectric material (Bi_2Te_3) on the silicon substrate and transferred thermoelectric material on epoxy, respectively.

Figure 3(a) and **(b)** show the electrodeposited thermoelectric materials by constant and pulsed conditions, respectively. As can be seen that, the constant electrodeposited film (**Figure 3(a)**) exhibits an initial 4 μm-thick compact layer while the top layer includes pillar structures. Although the thick-film thermoelectric material can be achieved by further deposition, its mechanical strength is very weak due to its porous structure. The thick electrodeposited film by the constant condition is easily peeled off for substrate. To overcome this problem, pulsed electrodeposition has been conducted. Compared with the constant electrochemical deposition, the pulsed electrodeposition with a pulse delay time for the recovery of the ion concentration always leads to a crystalline structure with high orientation and good uniformity [32]. This is proven in **Figure 3(b)**. The deposited surface under pulsed conditions is more uniform and smoother than that under constant conditions. **Figure 3(c)** shows a representative cross-sectional SEM image of the 600 μm-thick Bi_2Te_3 electrodeposited film, which is comparable to the bulk Bi_2Te_3 material. Consequently, by using simple and

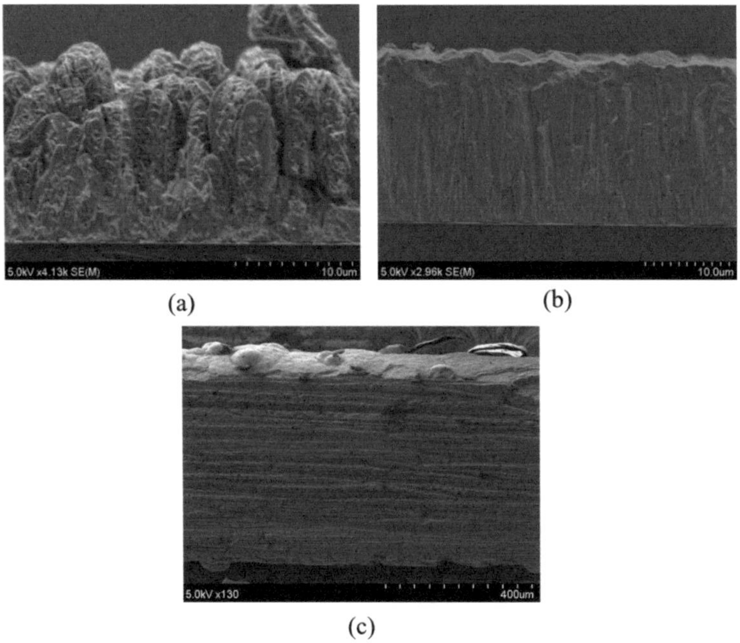

Figure 3.
Thermoelectric material. (a) Constant deposition. (b) Pulsed deposition. (c) A 600 μm-thick Bi_2Te_3 electrodeposited film.

	Constant electrodeposition		Pulsed electrodeposition	
	Nonannealing	**Annealing (250°C)**	**Nonannealing**	**Annealing (250°C)**
Seebeck coefficient (±20 μV/K)	−50	−110	−80	−150
Electrical resistivity (±5 μΩm)	50	20	20	15
Power factor (W/mK^2)	0.5×10^{-4}	6×10^{-4}	3.2×10^{-4}	15×10^{-4}

Table 1.
Electrodeposited thermoelectric material properties.

low-cost electrochemical deposition technique, thick bulk-like thermoelectric material posing a highly compact and uniform appearance could be achieved.

Thermoelectric material properties, including Seebeck coefficient and electrical resistivity, are evaluated, as shown in **Table 1**. The pulsed deposited film has a higher Seebeck coefficient as well as lower electrical resistivity than those of the constant deposited film. The power factor for pulsed deposited material is 3.2×10^{-4} W/mK^2 while it is 0.5×10^{-4} W/mK^2 for constant deposited material. Moreover, an annealing process has been performed to enhance the characteristics of the electrodeposited thermoelectric materials. The highest Seebeck coefficient is found at the annealing temperature of 250°C. The details of measurement setup and evaluation results can be found in [33].

In summary, thick bulk-like thermoelectric material based on the electrochemical deposition technique has been demonstrated. The electrodeposited film possesses a highly compact and uniform surface. The electrodeposited material properties by pulsed deposition are much higher than those by constant deposition. Also, thermoelectric performances of the electrodeposited film enhanced by the annealing process have been investigated.

3.3 Platinum nanoparticles embedded in thermoelectric material

Metal nanoparticle inclusion in the nanocomposite process is one of the promising methods to enhance the figure of merit *ZT*. However, there are a limited number of research studies on metal nanoparticle inclusion to improve thermoelectric material in film condition, especially through the synthesis of the electrochemical deposition. Au nanoparticle-Bi_2Te_3 nanocomposite has been demonstrated in [34], which is synthesized by a chemical-solution-based bottom-up method at low temperature. The ZT reaches up to 0.95 at 450 K [34]. A similar technique has been applied successfully for the Ag nanoparticle-Bi_2Te_3 nanocomposite, as shown in [35]. Nevertheless, its performance only improved significantly at a high-temperature region while at room temperature, its performance is just a half that of the pure Bi_2Te_3 because of the lower value of the Seebeck coefficient resulting in a smaller the ZT value. Herein, we select the Pt nanoparticles for embedding to Bi_2Te_3 because it has been proven by [36]. In this reference, the Pt nanoparticles have been embedded in Sb_2Te_3, which can enhance the Seebeck coefficient by filtering the low-energy carriers caused by band-bending potential formation, thus improving the power factor. Moreover, the Pt nanoparticles can help reduce the thermal conductivity due to scattering the mid- to long-wavelength phonons. Therefore, the ZT of nanocomposite thermoelectric material is much higher than that of pure thermoelectric material.

Figure 4.
(a) Electrodeposited surface of Bi_2Te_3. (b) Electrodeposited surface of Pt-Bi_2Te_3. (c) High resolution of TEM image of Pt-Bi_2Te_3.

Figure 4(a) shows the surface morphology of the electrodeposited pure Bi_2Te_3 with its crystal as plate-like structure. The surface morphology has been modified by the inclusion of Pt nanoparticles in the Bi_2Te_3, as shown in **Figure 4(b)**. The crystal grain size of Pt- Bi_2Te_3 composite is smaller than that of pure Bi_2Te_3, as can be seen in **Figure 4(a)** and **(b)**. Thus, the electrodeposited film with Pt nanoparticles tends to form lower porosity and denser surface structure in comparison to pure Bi_2Te_3. A high-resolution transmission electron microscopy image of Pt- Bi_2Te_3 composite is shown in **Figure 4(c)**, where black areas represent the Pt nanoparticles.

Table 2 shows the average grain size calculated by identifying FWHM and Integral Breadth β. As can be seen, the crystal's grain size becomes smaller at higher Pt nanoparticle content. The smallest grain size of 7.9 nm is found at the 1.9 wt% of Pt nanoparticles in the composite, which is four times smaller compared with that of pure Bi_2Te_3.

The summary of characteristic of the synthesized films is shown in **Table 3**. Experimental results indicate that once the grain size decreases, the carrier

Electrodeposited films	Deposited Pt (wt%)	Integral Breadth, β at $2\theta = 27.7°$, (rad)	Average grain size (nm)
Bi_2Te_3	0.0	0.6×10^{-2}	32.2 ± 4.3
Pt/Bi_2Te_3 -I	1.0	1.6×10^{-2}	13.9 ± 3.4
Pt/Bi_2Te_3 -II	1.5	2.2×10^{-2}	10.9 ± 1.3
Pt/Bi_2Te_3 -III	1.9	3.8×10^{-2}	7.9 ± 0.1

Table 2.
Average grain size on Bi_2Te_3 and Pt-Bi_2Te_3 nanocomposite films at $2\theta = 27.7°$.

Electrodeposited films	Average grain size (nm)	Electrical conductivity (S/cm)	Seebeck coefficient (μV/K)	Carrier concentration, n (cm^{-3}) × 10^{17}
Bi_2Te_3	36.5	618.7	−115.2	6.21
Pt (1.0 wt.%)/Bi_2Te_3	17.3	704.3	−152.1	2.40
Pt (1.5 wt.%)/Bi_2Te_3	12.1	643.7	−166.6	2.02
Pt (1.9 wt.%)/Bi_2Te_3	7.80	527.8	−184.1	1.93

Table 3.
Summary characteristics of the synthesized films.

concentration becomes lower. The lowest carrier concentration is observed for 1.9 wt% Pt-Bi_2Te_3 composite in comparison with others, including Bi_2Te_3, 1.5 wt% Pt-Bi_2Te_3, and 1.0 wt% Pt-Bi_2Te_3. As mentioned in Section 2, the Seebeck coefficient and electrical conductivity are trade-off, and they strongly depend on the carrier concentration. Lower carrier concentration results in a higher Seebeck coefficient but causes the smaller electrical conductivity, which agrees with the observation in this work, as given in **Table 3**.

Figure 5 shows the measurement result of the thermal conductivity of the electrodeposited film. The thermal conductivity decreases as the Pt nanoparticle concentration increases. The main reason is due to a reduction of the phonon mean free path caused by phonon grain boundary scattering [37]. The scattering mechanism of mid- to long-wavelength of phonons in the Pt-Bi_2Te_3 nanocomposite can be imagined via **Figure 5(b)**. Short-wavelength phonons are scattered by imperfections such as atomic defects and stacking defects while the Pt nanoparticles and grain boundaries are effective at scattering the mid-to long-wavelength phonon. A close adjacent between the Pt nanoparticles also contributed to the phonon scattering effect by reducing the phonon mean free path. Based on measurement results, including Seebeck coefficient, electrical conductivity, and thermal conductivity, the maximum ZT for Pt-Bi_2Te_3 nanocomposite is found at 0.61, which is 300% higher than that of the electrodeposited pure Bi_2Te_3. The details of evaluation setup, measurement results, and other discussions can be found in [24].

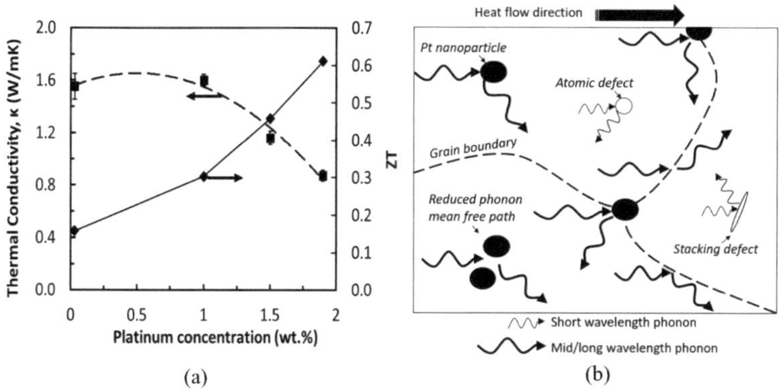

Figure 5.
Thermal conductivity and ZT as a function of Pt nanoparticle concentration. (b) Illumination of phonon scattering mechanisms in the Pt-Bi_2Te_3 nanocomposite.

In summary, Pt-Bi_2Te_3 nanocomposite has been synthesized successfully by the electrochemical deposition technique. It is found that as higher Pt nanoparticles are deposited in the nanocomposite film, the grain size becomes smaller and the nanostructure experienced significant defects. The change of grain size could be a help to adjust the trade-off between Seebeck coefficient and electrical conductivity, which results in the highest power factor. In addition, the defects caused by Pt nanoparticle benefit the phonon scattering enhancement, thus lowering the thermal conductivity. Consequently, the ZT can be improved.

3.4 Nickel-doped thermoelectric material

Although the thick-film thermoelectric materials have been investigated successfully, as described in Section 3.2, further investigations are still required to enhance their thermoelectric characteristics. Moreover, in order to open an opportunity for mass production, highly scalable synthesis electrodeposition on a large wafer size for thermoelectric materials should be conducted. In this section, a novel process technology for the ultra-thick film as well as high-performance characteristics (high Seebeck coefficient, large electrical conductivity, and low thermal conductivity) is investigated. Both electrodeposited films, including pure Bi_2Te_3 and Ni-doped Bi_2Te_3, reaching in mm-order thickness, have been synthesized, evaluated, and compared. Moreover, a highly scalable electrodeposition process for large wafer size has been performed and proven.

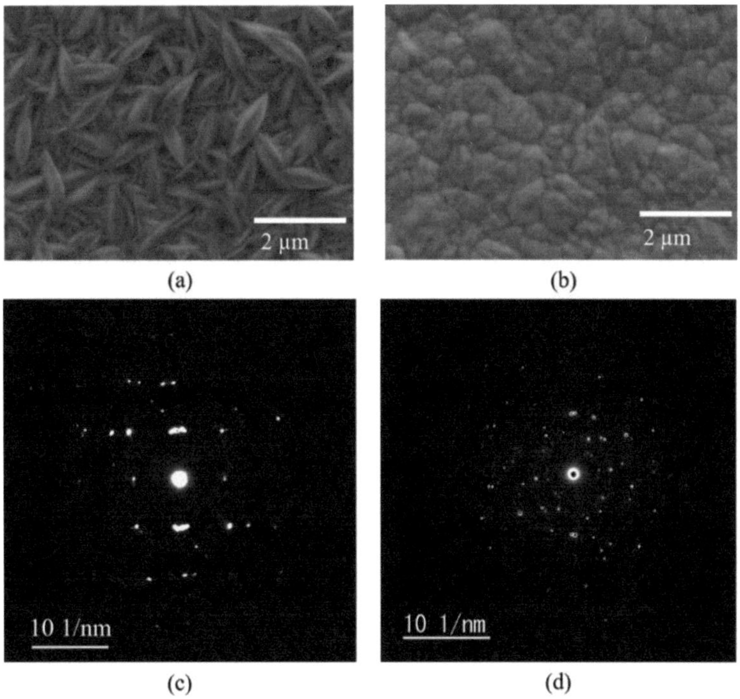

Figure 6.
SEM image of pure Bi_2Te_3. (b) SEM image of Ni doped Bi_2Te_3. (c) Selected area electron diffraction pattern of pure Bi_2Te_3. (c) Selected area electron diffraction pattern of pure Ni-doped Bi_2Te_3.

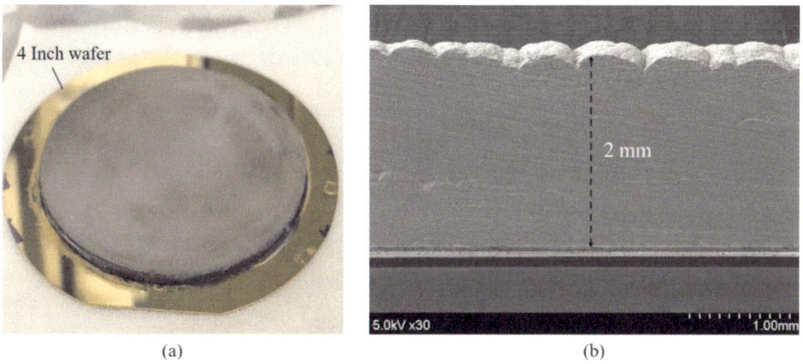

Figure 7.
(a) Electrodeposition on 4-inch wafer size. (b) SEM image of the cross-sectional view of the electrodeposited film.

Figure 6(a) and **(b)** show the surface crystal structure of the electrodeposited pure Bi_2Te_3 and Ni-doped Bi_2Te_3, respectively. As can be seen that the crystal grain size of pure Bi_2Te_3 is much larger than that of Ni-doped Bi_2Te_3. The selected area electrode diffraction patterns for pure Bi_2Te_3 and Ni-doped Bi_2Te_3 are shown in **Figure 6(c)** and **(d)**, respectively. Diffraction spots in **Figure 6(c)** and **(d)** indicate that both electrodeposited films pose polycrystalline structures. In quantitative comparison, the spots in **Figure 6(d)** are much more than those in **Figure 6(c)**. One possible cause is the grain size effects. Decreasing the grain size results in an increase of the boundary scattering and lattice defects, as discussed in Section 3.3. Thereby, not only the trade-off between Seebeck coefficient and electrical conductivity could be adjusted (changing the carrier concentration), but also the thermal conductivity gets lower due to photon scattering.

Figure 7(a) shows the experimental result of the highly scalable synthesis process, which is performed on a 4-inch wafer size. The deposited film reaches 2 mm thickness with a high uniform surface, as shown in **Figure 7(b)**. The success of the highly scalable electrodeposition could open up the opportunity for mass production to reduce the fabrication cost.

Summary characteristics of the electrodeposited thermoelectric materials can be found in **Table 4**. Experimental results show that 0.7 at% Ni-doped Bi_2Te_3 has the highest Seebeck coefficient as well as largest electrical conductivity compared with others, including pure Bi_2Te_3, 0.3 at% Ni-doped Bi_2Te_3, 1.0 at% Ni-doped Bi_2Te_3, and 1.5 at% Ni-doped Bi_2Te_3. Although the thermal conductivity of 0.7 at% Ni-doped Bi_2Te_3 is not the smallest one, its thermal conductivity is two times smaller than that of the pure Bi_2Te_3. The ZT of Ni-doped Bi_2Te_3 is estimated as 0.78, which is five times larger than that of the pure Bi_2Te_3. The details of evaluation setup, and measurement results, and other discussions can be found in [38, 39].

4. Device fabrication

4.1 Micro-thermoelectric-generator-based on micro/nano fabrication technology

One of the challenges for micro-TEG is the small harvested temperature difference across the module, thus resulting in low output power. In the conventional design of micro-TEG, the heat flows in the vertical direction (thermoelectric elements such as column structure); therefore, ultra-height thermoelectric elements are typically

	Seebeck coefficient (μV/K)	Electrical conductivity (S/cm)	Power factor (μV/m.K^2)	Thermal conductivity (W/m.K)	Figure of merit *ZT*
Pure Bi_2Te_3	−115 ± 5	525 ± 10	694	1.3 ± 0.1	0.15 ± 0.05
0.3 at% Ni-Bi_2Te_3	−130 ± 5	885 ± 30	1496	0.8 ± 0.05	0.61 ± 0.1
0.7 at% Ni-Bi_2Te_3	−143 ± 4	975± 15	2050	0.76 ± 0.09	0.78 ± 0.1
1.0 at% Ni-Bi_2Te_3	−125 ± 5	675± 70	1054	0.62 ± 0.04	0.52 ± 0.12
1.5 at% Ni-Bi_2Te_3	−130 ± 10	575 ± 75	972	0.56 ± 0.06	0.5 ± 0.18

Table 4.
Summary characteristics of the electrodeposited thermoelectric materials.

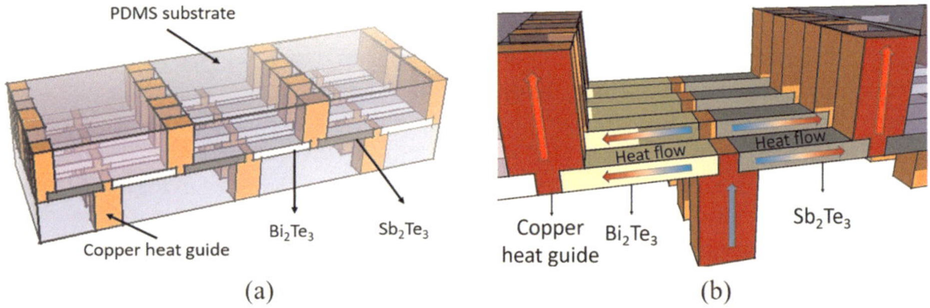

Figure 8.
(a) Proposed micro-thermoelectric generator structure. (b) Heat flow in lateral direction.

needed. However, to fabricate micro-TEG based on micro/nano technologies, the height of thermoelectric elements is limited to a hundred micrometers due to the limitation of the photoresist thickness and a patterning aspect ratio. To overcome this issue, thermoelectric elements are proposed to be laid in a lateral direction instead of a vertical one. The proposed structure for micro-TEG is shown in **Figure 8(a)**, which consists of n- and p-types thermoelectric elements (Bi_2Te_3 and Sb_2Te_3), copper heat guide, and PDMS (polydimethylsiloxane) as a base material. This micro-thermoelectric generator possesses a flexible characteristic that can be utilized in wearable electronic applications. The heat flow direction is shown in **Figure 8(b)**.

Figure 9 shows the fabrication process for micro-TEG, which begins with a silicon wafer. The SiO_2 with 500 nm thickness and Cr-Au layers with 10 nm thickness and 150 nm thickness, respectively, are deposited on the top of the silicon wafer, respectively, by PECVD and sputtering methods (**Figure 9(a)**). The thermoelectric materials are selectively deposited on the Au surface by electrodeposition technique via the patterned photoresist with a thickness of 100 μm (**Figure 9(b)**). Next, Ti-TiN-Au-Cu layer as a barrier contact layer is formed by sputter via a stencil mask, as shown in **Figure 9(c)–(e)**. The copper heat guides are subsequently grown on the barrier contact layer by the electroplating method (**Figure 9(f)**). The front side of micro-TEG is then filled by PDMS (**Figure 9(g)**). To create the heat guide from backside, a deep reactive ion etching (RIE) is conducted (**Figure 9(g)**). A thermal glue with high thermal conductivity is refilled into the molds by a screen printing technique (**Figure 9(h)**). The remaining silicon layer is etched out by plasma etching, and SiO_2

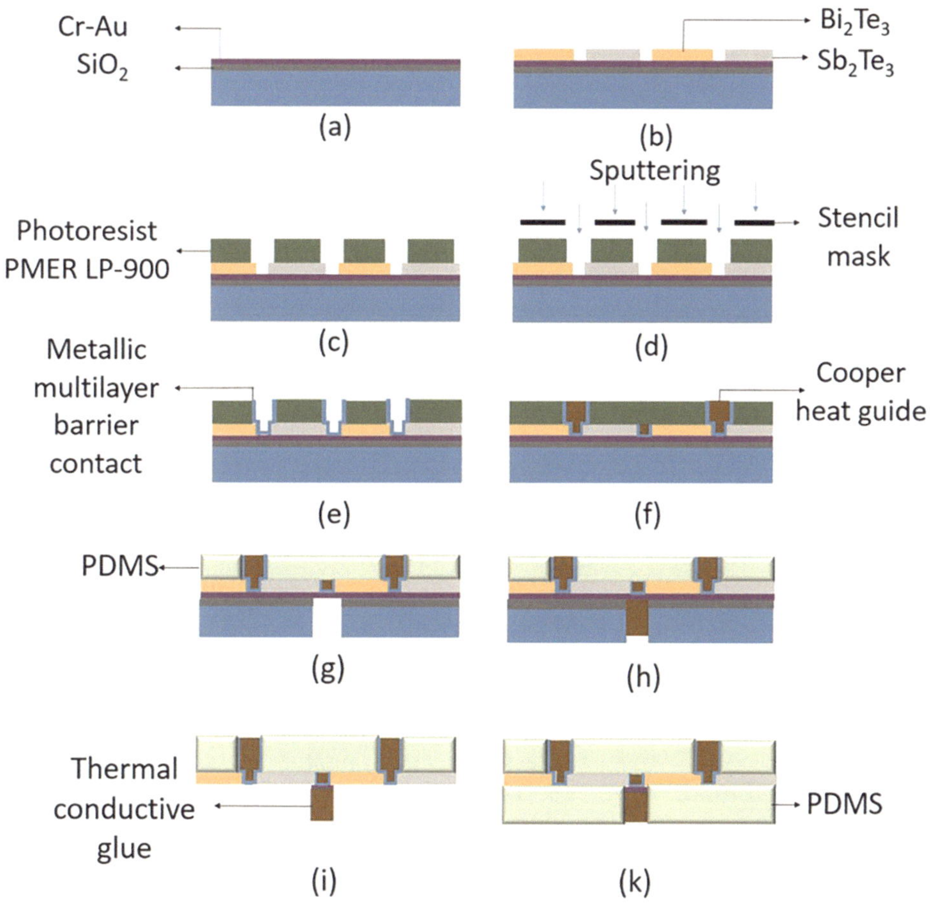

Figure 9.
Fabrication process. (a) SiO_2-Cr-Au deposition. (b) Thermoelectric material synthesis. (c) Photolithography process. (d, e) Multilayers of barrier metal contacts of Ti-TiN-Au-Cu. (f) Copper heat guides. (g) PDMS refilling and Si-SiO_2 removing processes; (h) screen printing process of thermal conductive glue. (i) Backside etching process; (k) PDMS refilling process.

and Cr-Au layers are removed by the ion beam milling technique (**Figure 9(i)**). Finally, PDMS is filled into the backside cavities (**Figure 9(k)**).

Figure 10(a) shows the fabricated micro-TEG based on micro/nano fabrication technologies. The micro-TEG contains 24 pairs of electrodeposited n- and p-type thermoelectric materials integrated on 1 cm^2. The output power density of the fabricated micro-TEG is displayed in **Figure 10(b)**, which reaches 3 $\mu W/cm^2$ under a temperature difference caused by human body (37°C) and ambient environment (15°C) using natural convection. The details of evaluation setup, measurement results, and other discussions can be found in [40].

In summary, a novel design and fabrication process for the micro-TEG have been proposed and investigated. Micro-TEG has been fabricated successfully by micro/nano fabrication technologies. Also, its performance has been evaluated. Although the power density of the fabricated micro-TEG is small, it could be improved by increasing the density of n- and p-types thermoelectric elements. The idea and experimental results in this work may be useful for applications in wearable electronic devices.

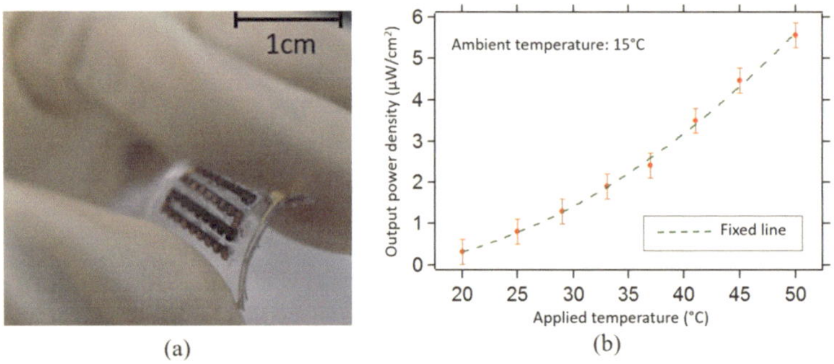

Figure 10.
(a) Fabricated micro-TEG. (b) Applied temperature and output power.

4.2 Micro-thermoelectric generator based on assembling technology

To improve the performance of the micro-TEG, enhancing the performance of the thermoelectric materials is a critical point. Another important point is an increase in the number of thermoelectric elements, which can significantly enhance output voltage and output power, as discussed by Eqs. (12) and (13). Thus, the power density can be significantly increased. High-density n- and p-type thermoelectric elements could be formed on a small foot print by utilizing the micro/nano fabrication technologies, as discussed in Section 4.1 and in Refs. [41, 42]. However, some issues need to be addressed, as follows. Complex processes, including photolithography, etching, deposition, and lift-off processes, are needed to construct the air bridge between thermoelectric elements. Therefore, the fabrication time is long, and the cost is high. Moreover, the bonding strength between thermoelectric elements and substrate is weak; thereby, the internal resistance of the fabricated micro-TEG is high, caused by the large contact resistance. Such issues make the performance of the micro-TEG low, which is against it for realistic applications. In this section, a novel method to produce the micro-TEG based on ultra-thick and dense electrodeposited thermoelectric material (presented in Section 3.4) and assembly technique is proposed and investigated.

To fabricate a high-density micro-TEG, small thermoelectric elements are needed, which are prepared as follows. The 4-inch electrodeposited wafer (**Figure 11(a)**) is diced into many small elements (**Figure 11(b)**). It is noted that before cutting, Ni-Au

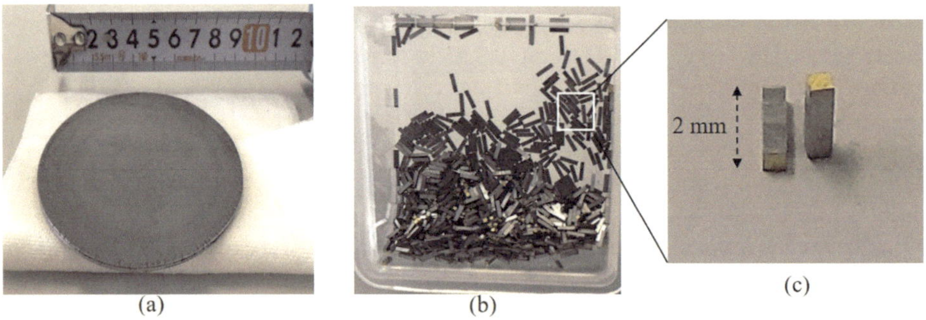

Figure 11.
(a) Four-inch electrodeposited thermoelectric material wafer. (b) Thermoelectric elements with dimensions of 0.4 mm × 0.4 mm × 2 mm. (e) Close-up image of thermoelectric elements.

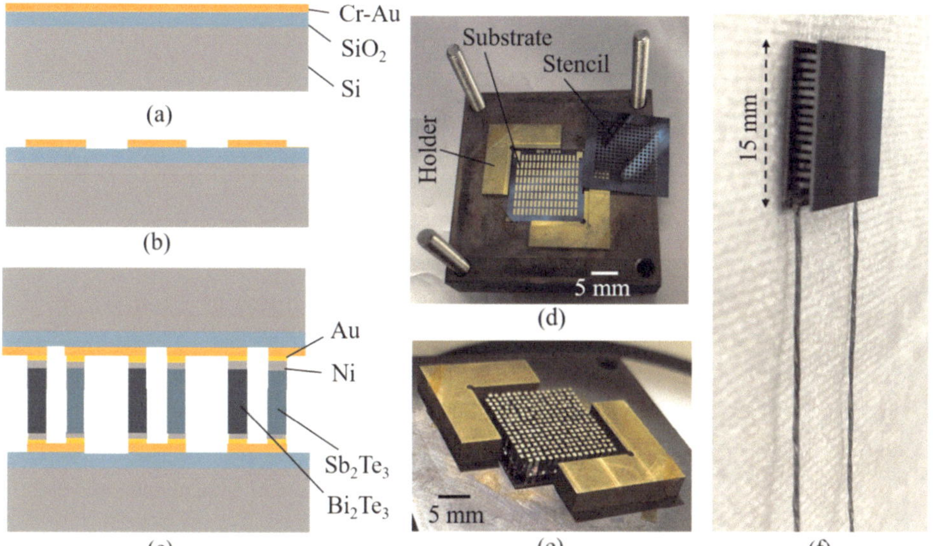

Figure 12.
Fabrication process and fabricated micro-TEG. (a) Silicon substrate with SiO_2 and Cr-Au layers on top. (b) Cr-Au patterning. (c) TEG schematic. (d) Device fabrication setup including holders, substrate, and stencil wafer. (e) After the first alignment and bonding. (f) Completely fabricated device.

layers as barrier contact layers are formed on both sides of the wafer by electroplating method [43, 44] to decrease the ohmic contact resistance between thermoelectric elements and substrate. **Figure 11(c)** shows the magnified image of the diced thermoelectric elements with dimensions of 0.4 mm × 0.4 mm × 2 mm.

The fabrication process for the micro-TEG based on the assembly technique is shown in **Figure 12(a)–(c)**. The SiO_2 layer as an insulator layer is formed on a silicon wafer by PECVD, and Cr-Au layers are deposited on the SiO_2 layer by the sputtering method, as given in **Figure 12(a)**. Cr-Au layers are patterned to form the bottom interconnection by a wet etching method [45, 46], as shown in **Figure 12(b)**. Next, thermoelectric elements are aligned and bonded on the substrate by conductive glue. Finally, a top wafer cover is aligned and bonded on top of the thermoelectric elements (**Figure 12(c)**). Because the thermoelectric elements are pretty small, the process for vertical alignment becomes difficult. To overcome this issue, a stencil silicon wafer with patterned through holes is proposed, and a simple metal holder tool is employed to fix and align the stencil wafer and substrate, as shown in **Figure 12(d)**. Thermoelectric elements are inserted into holes of the stencil wafer. **Figure 12(e)** shows the experimental image after the thermoelectric elements are bonded on the substrate. The completely fabricated micro-TEG is shown in **Figure 12(f)**. In total, 127 pairs, including n- and p-type thermoelectric elements, are formed successfully on a small footprint of 15 mm^2. Thus, although a simple assembly technique is employed, the integration density of thermoelectric elements could be comparable to the micro-fabrication of the micro-TEG.

The fabricated micro-TEG shows a high output power of 33.9 mW and a large power density of 15.1 mW/cm^2 under a temperature difference across the micro-TEG of 75 °C, which is much higher performance than those of other published works [42, 47–51]. More comparisons to other works are shown in **Table 5**. The details of evaluation setup, measurement results, and other discussions can be found in [52].

Pairs	Height (mm)	Temperature difference ΔT (°C)	Open circuit (V)	Internal resistance (Ω)	Power (mW)	Power density (mW/cm^2)	References
24	0.2	24	0.05	200	—	0.004	[47]
6	0.0015	6	0.036	25	0.0023	—	[48]
71	0.0135	39	0.2	134	2.4	2.4	[49]
127	0.01	52.5	0.3	13	3	9.2	[50]
200	0.02	88	0.5	45.2	—	1.04	[42]
220	2	40	2.1	—	7	1.75	[51]
127	2	75	2.2	35	33.9	15.1	This work [52]

Table 5.
Comparison of TEG performance.

In summary, the high integration density of the micro-TEG has been demonstrated by utilizing a simple assembly technique. Micro-TEG consisting of 127 pairs is successfully fabricated on 15 mm^2. The fabricated micro-TEG possesses a high performance, which may satisfy the demand for being a reliable power source for electronic devices.

5. Application demonstration

Although a high output voltage and output power could be achieved by the fabricated micro-TEG, a high thermal source is needed. In turn to low-thermal sources, its output power is in small value, which cannot be used as a power source for electronic devices. To overcome this issue, a DC-DC converter is required, which amplifies the output voltage of the micro-TEG from an mV range to V range of the output of the DC-DC converter. Thus, this makes micro-TEG possible for powering electronic devices with low-power consumption. In this section, the micro-TEG for powering calculator and twist watch is demonstrated. A DC-DC converter is utilized to boost the output voltage of the micro-TEG up to sufficient levels to store in an energy-storable unit, which is subsequently supplied to electronic devices. The energy storable unit can be a capacitor, a supercapacitor, or a rechargeable battery. We have developed successfully micro-supercapacitors-based graphene nanowalls with PANI in liquid state [53] and solid state [54] and with MnO_2 [55]. Although these micro-supercapacitors show a high charge and discharge processes, their storable energy is lower than that of commercial rechargeable battery. In this section, a rechargeable battery from Enercera [56] is employed for the application demonstration. Two applications utilizing the micro-TEG are conducted, as follows.

5.1 Micro-TEG for powering portable electronic devices

Figure 13(a) illustrates the experimental setup for the micro-TEG as a power source for the calculator. It consists of Peltier (as a heat source), copper blocks, temperature sensors, the DC-DC converter, a rechargeable battery, and a calculator.

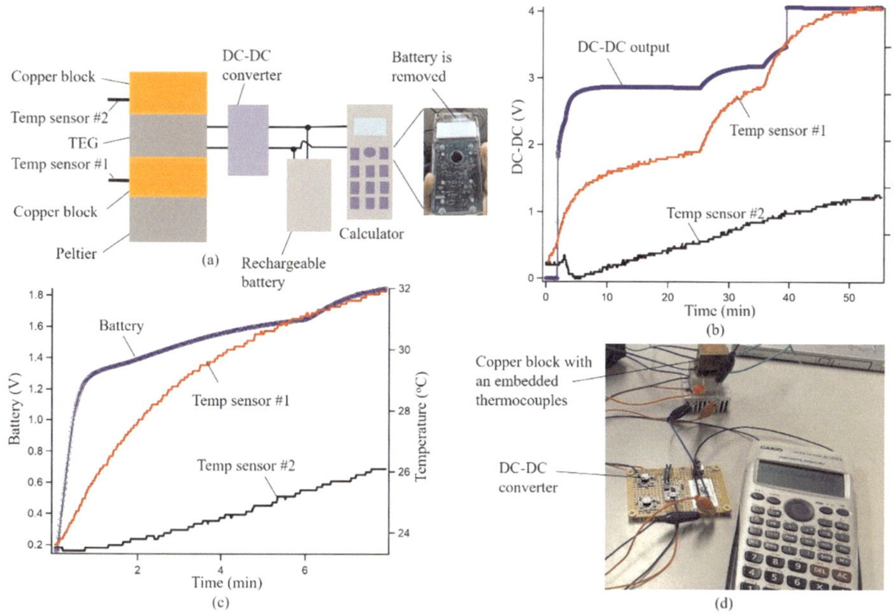

Figure 13.
(a) Experimental setup for powering portable electronic device. (b) DC-DC output as a function of temperature difference. (c) Battery charged up by the micro-TEG. (d) Micro-TEG as a power source for calculator.

The harvester energy is accumulated and stored in the rechargeable battery via the DC-DC converter and then supplied to electronic devices. **Figure 13(b)** shows the output of DC-DC converter over the temperature difference across the micro-TEG. The experimental results indicated that output of DC-DC converter reaches 2.8 V at $\Delta T = 2°C$ and 4 V at $T = 8°C$. **Figure 13(c)** shows the rechargeable battery characteristic, which increases from 0 V to 1.8 V, taking approximately 8 minutes. **Figure 13(d)** shows the demonstration of using micro-TEG as an electrical power source for the calculator. The calculator can be powered on and used once the rechargeable battery gets over 1.5 V.

5.2 Micro-TEG for powering wearable electronic devices

Figure 14(a) illustrates the experimental setup for powering a twist watch. One side of the micro-TEG is in contact with human skin while another side is attached to the backside of the twist watch. α-Gel is pasted on both sides of the micro-TEG to enhance heat transfer between interfaces. The DC-DC converter and rechargeable battery are employed, which are similar to those mentioned in Section 5.1. The DC-DC converter, rechargeable battery, and micro-TEG are arranged on the twist watch, as shown in **Figure 14(b)**. **Figure 14(c)** shows the output of the micro-TEG and battery charge when twist watch is worn. It takes approximately 5 minutes for the rechargeable battery to reach 1.2 V. With this energy, the twist watch is powered on and runs.

Demonstrated results in this section indicate a high potential using the micro-TEG for powering not only portable electronic devices but also wearable electronic devices. Further integrated functions, including sensing (humidity, temperature, gases, etc.),

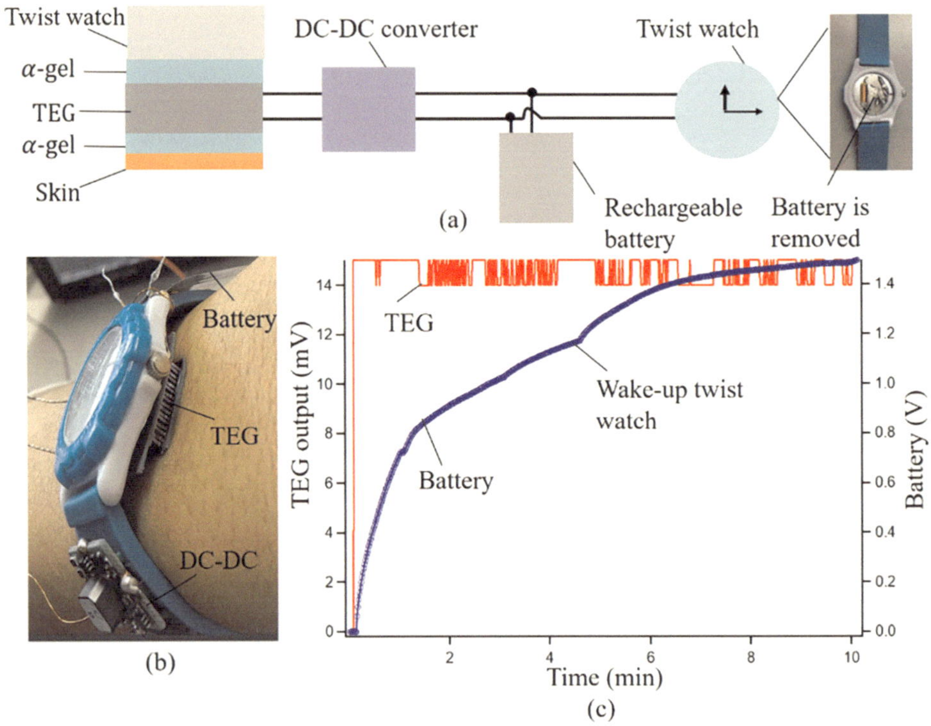

Figure 14.
(a) Experimental setup for powering wearable electronic device. (b) The photo of the self-powered twist watch. (c) TEG output and battery charge-up.

displaying (screen display), and transmitting (radio frequency, Bluetooth, etc.) functions, should be investigated to produce a smart system for using in wireless IoT sensing systems.

6. Conclusions

In this work, not only basic knowledge about thermoelectric generators but also experiences on material synthesis, device fabrication, and application demonstration are reported. By investigating electrochemical deposition, high-performance thermoelectric materials have been achieved. Three kinds of high-performance thermoelectric materials, including thick bulk-like thermoelectric material, Pt nanoparticles embedded in a thermoelectric material, and Ni-doped thermoelectric material, are reported and discussed. Besides the material synthesis, novel fabrication methods can also help increase the output power and the power density of the micro-TEG significantly. Two fabrication processes, micro/nano fabrication technology and assembly technology, are investigated to produce high-performance micro-TEG. Moreover, the fabricated micro-TEG is successfully demonstrated for powering portable and wearable electronic devices. The contents of this paper are based on our experimental research. It is our hope that this review may be a useful reference for those working in the field of thermal-to-electric energy conversion, especially on the micro-TEG.

Acknowledgements

Part of this work was performed in the Micro/Nanomachining Research Education Center (MNC) of Tohoku University. This work was supported by Cabinet Office, Government of Japan, Cross-ministerial Strategic Innovation Promotion Program (SIP), (funding agency: The New Energy and Industrial Technology Development Organization, NEDO) and also supported in part by JSPS KAKENHI for Young Scientists (Grant number: 20K15147).

Author details

Nguyen Van Toan[1]*, Truong Thi Kim Tuoi[2], Nguyen Huu Trung[3], Khairul Fadzli Samat[4], Nguyen Van Hieu[5] and Takahito Ono[1,2]*

1 Micro System Integration Center, Tohoku University, Sendai, Japan

2 Graduate School of Engineering, Tohoku University, Sendai, Japan

3 National Institute of Industrial Science and Technology, Nagoya, Japan

4 Fakulti Kejuruteraan Pembuatan, Universiti Teknikal Malaysia Melak, Melaka, Malaysia

5 Department of Physics and Electronic Engineering, University of Science, VNU-HCM, Vietnam

*Address all correspondence to: nguyen.van.toan.c6@tohoku.ac.jp; takahito.ono.d4@tohoku.ac.jp

References

[1] Chang SY, Cheng P, Li G, Yang Y. Transparent polymer photovoltaics for solar energy harvesting and beyond. Joule. 2018;**2**:1039-1054

[2] Tran LG, Cha HK, Park WT. RF power harvesting: A review on designing methodologies and applications. Micro and Nano Systems Letters. 2017;**5**:14

[3] Toan NV, Hasana MMIM, Udagawa D, Inomata N, Toda M, Said SM, et al. Thermoelectric generator battery using 10 nm diameter of Al_2O_3 nanochannels for low-grade waste heat energy harvesting. Energy Conversion and Management. 2019;**199**:111979

[4] Tuoi TTK, Toan NV, Ono T. Theoretical and experimental investigation of a thermoelectric generator (TEG) integrated with a phase change material (PCM) for harvesting energy from ambient temperature changes. Energy Reports. 2020;**6**: 2022-2029

[5] Haras M, Skotnicki T. Thermoelectricity for IoT – A review. Nano Energy. 2018;**54**:461-476

[6] Tuoi TTK, Toan NV, Ono T. Heat storage thermoelectric generator as an electrical power source for wireless IoT sensing systems. International Journal of Energy Research. 2021;**45**:15557-15568

[7] Zhou Y, Liu Y, Zhou X, Gao Y, Gao C, Wang L. High performance p-type organic thermoelectric materials based on metalloporphyrin/single-walled carbon nanotube composite films. Journal of Power Sources. 2019;**423**: 152-158

[8] Xu S, Liu C, Xia Z, Zhong W, Luo Y, Ou H, et al. Cooperative effect of carbon black and dimethyl sulfoxide on PEDOT: PSS hole transport layer for inverted planar perovskite solar cells. Solar Energy. 2017;**158**:125-132

[9] Toshima N, Jiravanichanun N, Marutani H. Organic thermoelectric materials composed of conducting polymers and metal nanoparticles. Journal of Electronic Materials. 2012;**41**: 1735-1742

[10] Li Y, Toan NV, Wang Z, Samat KF, Ono T. Thermoelectrical properties of silicon substrates with nanopores synthesized by metal-assisted chemical etching. Nanotechnology. 2020;**31**: 455805

[11] Sabran NH, Fadzallah IA, Ono T, Said SM, Sabri MFM. Preparation and characterization of electrochemical deposition cobalt triantimonide (CoSb3) thick film: Effect of polyvinyl alcohol (PVA) as an additive. Journal of Electrochemical Materials. 2019;**48**: 5003-5011

[12] Tuoi TTK, Toan NV, Ono T. Heat storage thermoelectric generator for wireless IoT sensing systems. In: The 21st International Conference on Solid-State Sensors, Actuators and Microsystems (Transducers). Orlando, FL, USA: IEEE; 2021. pp. 924-927. DOI: 10.1109/Transducers50396.2021.9495686

[13] Zhang Y, Xing C, Zhang T, Li M, Pacios M, Yu X, et al. Tin selenide molecular precursor for the solution processing of thermoelectric materials and devices. ACS Applied Materials & Interfaces. 2020;**12**: 27104-27111

[14] Ono T, Nguyen TH, Samat KF, Li J, Toan NV. Nanoengineered thermoelectric energy devices for IoT

sensing applications. ECS Transactions. 2019;**92**:163-168

[15] Dheepa J, Sathyamoorthy R, Velumani S, Subbarayan A, Natarajan K, Sebastian PJ. Electrical resistivity of thermally evaporated bismuth telluride thin films. Solar Energy Materials and Solar Cells. 2004;**81**:305-312

[16] Bendt G, Gassa S, Rieger F, Jooss C, Schulz S. Low-temperature MOCVD deposition of Bi_2Te_3 thin films using Et2BiTeEt as single source precursor. Journal of Crystal Growth. 2018;**490**: 77-83

[17] Carter MJ, Desouky AE, Andre MA, Bardet P, Leblance S. Pulsed laser melting of bismuth telluride thermoelectric materials. Journal of Manufacturing Processes. 2019;**43**:35-46

[18] Rosi FD. Thermoelectricity and thermoelectric power generator. Solid State Electronics. 1968;**11**:833-868

[19] Heremans JP, Jovovicm V, Toberer ES, Saramat A, Kurosaki K, Charoenphakdee A, et al. Enhancement of thermoelectric of the electronic density of states. Science. 2008;**321**: 1457-1461

[20] Pei Y, Shi X, Lalonde A, Wang H, Chen L, Snyder GJ. Convergence of electronic bands for high performance bulk thermoelectrics. Nature. 2011;**473**: 66-69

[21] Rosi FD. Thermoelectricity and thermoelectric power generation. Solid-State Electronics. 1968;**11**:833-848

[22] Chasmar RP, Stratton R. The thermoelectric figure of merit and its relation to thermoelectric generators. Journal of Electronic and Control. 1959;7: 52-57

[23] Hodes M. Optimal pellet geometries for thermoelectric power generation. IEEE Transactions on Components and Packaging. 2010;**33**:307-318

[24] Samat KF, Trung NH, Ono T. Enhancement in thermoelectric performance of electrochemically deposited platinum-bismuth telluride nanocomposite. Electrochimica Acta. 2019;**312**:62-71

[25] Li Y, Toan NV, Wang Z, Samat KF, Ono T. Formation and evaluation of silicon substrate with highly-doped porous Si layers formed by metal-assisted chemical etching. Nanoscale Research Letters. 2021;**16**:64

[26] Samat KF, Li Y, Toan NV, Ono T. Carbon black nanoparticles inclusion in bismuth telluride film for micro thermoelectric generator application. In: The 33rd IEEE International Conference on Micro Electro Mechanical Systems. Vancouver, BC, Canada: IEEE; 2020. pp. 562-565

[27] Kim C, Yang Y, Baek JY, Lopez DH, Kim DH, Kim H. Concurrent defects of intrinsic tellurium and extrinsic silver in an n-type Bi2te2.88Se0.15 thermoelectric materials. Nano Energy. 2019;**60**:26-35

[28] Ming T, Yang W, Huang X, Wu Y, Li X, Liu J. Analytical and numerical investigation on a new compact thermoelectric generator. Energy Conversion and Management. 2017;**132**: 261-271

[29] Macia E. Thermoelectric Materials: Advances and Applications. New York: Stanford Publishing; 2015. p. 364

[30] Kanani N. Chapter 5 – Electrodeposition considered at the atomistic level. In: Electroplating. Elsevier; 2004. pp. 141-177. DOI: 10.1016/B978-185617451-0/50005-1

[31] Bicer M, Sisman I. Electrodeposition and growth mechanism of SnSe thin films. Applied Surface Science. 2011;**257**: 2944-2949

[32] Ma Y. Thermoelectric characteristics of electrochemically deposited Bi_2Te_3 and Sb_2Te_3 thin films of relevance to multilayer preparation. Journal of the Electrochemical Society. 2012;**159**:50

[33] Trung NH, Sakamoto K, Toan NV, Ono T. Synthesis and evaluation of thick films of electrochemically deposited Bi_2Te_3 and Sb_2Te_3 thermoelectric materials. Materials. 2017;**10**:154

[34] Lee E, Ko J, Kim JY, Seo WS, Choi SM, Lee KH, et al. Enhanced thermoelectric properties of Au nanodot-included Bi_2Te_3 nanotube composites. Journal of Materials Chemistry C. 2016;**4**(6):1313-1319

[35] Zhang Q, Ai X, Wang L, Chang Y, Luo W, Jiang W, et al. Improved thermoelectric performance of silver nanoparticles-dispersed Bi_2Te_3 composites deriving from hierarchical two-phased heterostructure. Advanced Functional Materials. 2015;**25**(6): 966-976

[36] Sun T, Samani MK, Khosravian N, Ang KM, Yan Q, Tay BK, et al. Enhanced thermoelectric properties of n-type $Bi_2Te_{2.7}Se_{0.3}$ thin films through the introduction of Pt nanoinclusions by pulsed laser deposition. Nano Energy. 2014;**8**:223-230

[37] Touzelbaev MN, Zhou P, Venkatasubramanian R, Goodson KE. Thermal characterization of Bi2Te3/Sb2Te3 superlattices. Journal of Applied Physics. 2001;**90**:763

[38] Toan NV, Tuoi TTK, Ono T. Thermoelectric generators for heat harvesting: From material synthesis to device fabrication. Energy Conversion and Management. 2020;**225**:113442

[39] Toan NV, Tuoi TTK, Samat KF, Sui H, Inomata N, Toda M, et al. High performance micro-thermoelectric generator based on metal doped electrochemical deposition. In: The 33rd IEEE International Conference on Micro Electro Mechanical Systems. Vancouver, BC, Canada: IEEE; 2020. pp. 570-573

[40] Trung NH, Toan NV, Ono T. Electrochemical deposition based flexible thermal electric power generator with Y-type structure. Applied Energy. 2018;**210**:467-476

[41] Synder GJ, Jim JR, Huang CK, Fleurial JP. Thermoelectric microdevice fabricated by a MEMS-like electrochemical process. Nature Materials. 2003;**2**:528-531

[42] Yu Y, Zhu W, Wang Y, Zhu P, Peng K, Deng Y. Towards high integration and power density: Zigzag-type thin-film thermoelectric generator assisted by rapid pulse laser patterning technique. Applied Energy. 2020;**285**: 115404

[43] Toan NV, Sangu S, Ono T. Fabrication of deep SiO_2 and Tempax glass pillar structures by reactive ion etching for optical modulator. Journal of Microelectromechanical Systems. 2016; **25**:668-674

[44] Toan NV, Sangu S, Ono T. High aspect ratio SiO_2 pillar structures capable of the integration of an image sensor for application of optical modulator. IEEJ Transactions on Sensors and Micromachines. 2016;**136**:41-42

[45] Toan NV, Toda M, Ono T. An investigation on etching techniques for glass micromachining. Micromachining. 2016;**7**:51

[46] Toan NV, Ono T. Progress in performance enhancement methods for capacitive silicon resonators. Japanese Journal of Applied Physics. 2017;**56**: 110101

[47] Trung NH, Toan NV, Ono T. Fabrication of π-type flexible thermoelectric generators using an electrochemical deposition method for thermal energy harvesting applications at room temperature. Journal of Micromechanics and Microengineering. 2017;**27**:125006

[48] Shen H, Lee H, Han S. Optimization and fabrication of a planar thermoelectric generator for a high-performance solar thermoelectric generator. Current Applied Physics. 2021;**22**:6-13

[49] Roth R, Rostek R, Cobry K, Kohler C, Groh M, Woias P. Design and characterization of micro thermoelectric cross-plane generators with electroplated Bi_2Te_3, SbxTey, and reflow soldering. Journal of Microelectrochanical Systems. 2014;**23**: 961-971

[50] Zhang W, Yang J, Xu D. A high power density micro-thermoelectric generator fabricated by an integrated bottom-up approach. Journal of Microelectromechanical Systems. 2016; **25**:744-749

[51] Lee B, Cho H, Park KT, Kim JS, Park M, Kim H, et al. High-performance compliant thermoelectric generators with magnetically self-assembled soft heat conductors for self-powered wearable electronics. Nature Communications. 2020;**11**:5948

[52] Toan NV, Tuoi TTK, Hieu NV, Ono T. Thermoelectric generator with a high integration density for portable and wearable self-powered electronic devices. Energy Conversion and Management. 2021;**245**:114571

[53] Li J, Toan NV, Wang Z, Ono T. Metal-assisted chemical etching of silicon nanowires for templating 3D graphene growth towards energy storage in microsystems. Journal of Micromechanics and Microengineering. 2019;**29**:055077

[54] Toan NV, Tuoi TTK, Li J, Inomata N, Ono T. Liquid and solid states on-chip micro-supercapacitors using silicon nanowire-graphene nanowall-PANI electrode based on microfabrication technology. Materials Research Bulletin. 2020;**131**:110977

[55] Sui H, Toan NV, Ono T. Vertically-oriented graphene electrodeposited with MnO2 on native SiO2/Si for high-performance supercapacitor electrodes. Journal of Electroanalytical Chemistry. 2021;**895**:115507

[56] Available from: https://www.ngk-insulators.com/en/product/enercera.html [Accessed: January 12, 2022]

Chapter 3

Perspective Chapter: Ultra-Low Temperature Chillers for Semiconductor Manufacturing Process

Jung-In Yoon, Chang-Hyo Son, Sung-Hoon Seol and Ji-Hoon Yoon

Abstract

The growth of the semiconductor market and advancement of manufacturing technology have led to an increase in wafer size and highly integrated semiconductor devices. The temperature of the supplied cooling medium from the chiller that removes the heat produced in the semiconductor manufacturing process is required to be at a lower level because of the high integration. The Joule-Thomson cooling cycle, which uses a mixed refrigerant (MR) to produce the cooling medium at a level of −100°C required for the semiconductor process, has recently gained attention. When a MR is used, the chiller's performance is heavily influenced by the composition and proportions of the refrigerant charged to the chiller system. Therefore, this paper introduces a cooling cycle that uses an MR to achieve the required low temperature of −100°C in the semiconductor manufacturing process and provides the results of simple experiments to determine the effects of different MR compositions.

Keywords: Semiconductor etching process, Mixed refrigerant refrigerator, Refrigerant mixing ratio, Ultra low temperature, Joule-Thomson cycle

1. Introduction

Because of the growth of the semiconductor market and the global competition among many companies of the United States, Taiwan, South Korea, etc., investment and interest in related industries are increasing [1]. As a result of this trend, there is an increased demand for chillers, which are temperature control systems used in the semiconductor manufacturing process, and the research into chillers is also progressing actively [2].

As shown in **Figure 1** [3], semiconductor manufacturing process consists of eight major processes, including wafer manufacturing, oxidation, photolithography, etching, and deposition [4]. Etching, one of the eight major semiconductor processes, is a process that removes unnecessary parts in the sketch of circuit drawn in the

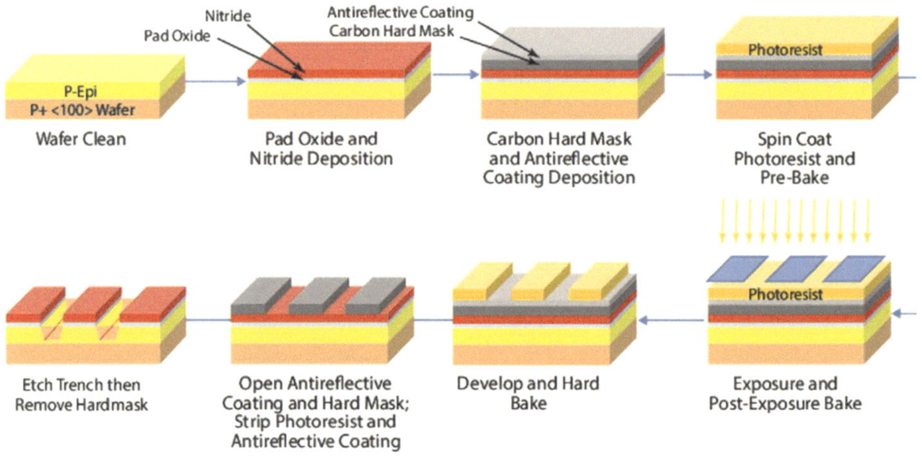

Figure 1.
Manufacturing process of semiconductors [3].

photolithography process [5]. A semiconductor etcher is a device used to etch circuit patterns formed on wafers. The typical method is to create plasma gas with excellent etching properties and etch a specified part of the wafer [6, 7].
Here, there might be a risk of an excessive rise in wafer temperature when it is exposed to the high temperature of the plasma [8]. Therefore, general etching devices cool the wafers by circulating the cooling medium inside the electrostatic chuck (ESC) where the wafers are installed [9].

On the contrary, because of the high integration of semiconductor circuit patterns, the line width of circuits is becoming smaller [10]. The temperature of ESC in the 20 ~ 30 nm line-width etching process is room temperature, but the −20°C temperature level is primarily used at the 10 nm level. The etching processes that require a lower temperature are trending especially at the fine level of 5 ~ 7 nm process. Therefore, there is a growing demand for the ultra-low temperature etching process, which involves performing etching while maintaining the temperature of substrates below −100°C. If etching is performed in the ultra-low temperature domain, the spontaneous reaction is suppressed, enabling anisotropic etching. However, this kind of process has disadvantages in that implementation is difficult in terms of the equipment and environment for maintaining the ultra-low temperature of substrates, and the energy consumption is high.

To achieve the low temperature of −100°C, various cycles such as, two-stage cascade cycle, Joule-Thomson cycle, and auto-cascade cycle are used. Joule-Thomson coolers are used in many fields because they have a simpler structure and are easier to manufacture and operate compared with the two-stage cascade cycle. However, the main disadvantage of Joule-Thomson coolers is their low efficiency, which is caused by irreversibility because of the high-pressure rate and wide temperature range. This problem can be resolved by using a mixed refrigerant (MR). **Figure 2** shows the refrigeration effect of a single cooling medium according to the working pressure. Nitrogen or argon, which is a low boiling point cooling medium, requires a higher pressure to obtain the cooling calories compared to propane or ethane, which is a high boiling point cooling medium. In the case of nitrogen, for example, a working pressure of approximately 24.6 MPa is required to obtain a cooling calorie of 1,000 J/mol, whereas, in the case of propane, the working pressure may be at a level of approximately 0.79 MPa, showing a large difference [11]. In fact, if an MR is created by combining

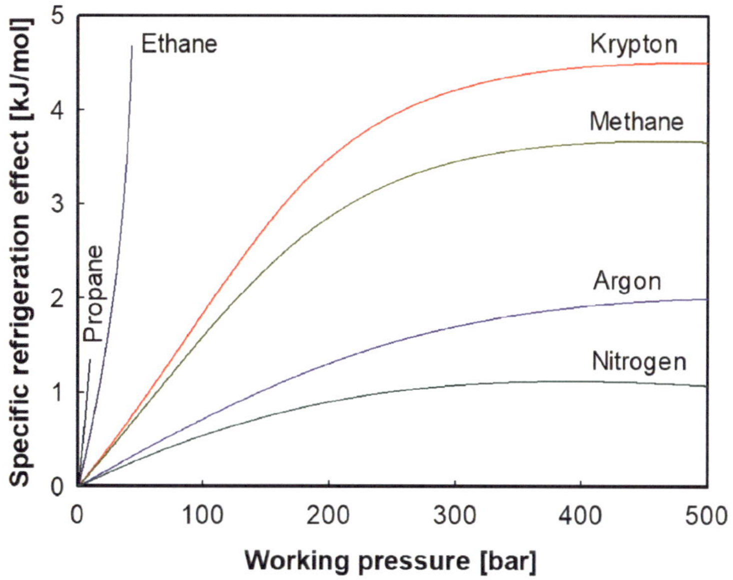

Figure 2.
Variation of specific refrigeration effect according to operation pressure [11].

nitrogen, argon, methane, krypton, ethane, propane, etc., the cooling calorie of 1,000 J/mol can be obtained below the working pressure of 2.5 MPa.

Therefore, this paper introduces a Joule-Thomson cooler that uses an MR to achieve the low temperature of −100°C required in the semiconductor manufacturing process.

2. Refrigeration cycles for ultra-low temperature

As aforementioned, a two-staged cascade cycle, Joule-Thomson cycle, etc. are used to achieve the low temperature of −100°C. This section introduces several refrigeration systems that are commonly used to achieve an ultra-low temperature [12].

2.1 Single mixed refrigerant (MR) refrigerator

A single MR refrigerator is a refrigerator that can achieve temperatures of −100°C or lower and has the advantage of having fewer mechanical elements, and it can be miniaturized. Because of these advantages, single MR refrigerators have been widely used in the semiconductor industry, which requires low-temperature refrigerators that can be operated reliably for an extended period. A single MR refrigerator has a single-stage refrigeration cycle that includes an intermediate heat exchanger, and it is configured as shown in **Figure 3**.

The specific working principle of single MR refrigerators is as follows. The refrigerant vapor sucked by the compressor flows into the aftercooler in a state of high temperature and high pressure through the compression process (1–2a). The refrigerant vapor is then cooled to ambient temperature through heat exchange with the heat exchanging medium at room temperature (2a–2), and flows into the intermediate heat exchanger.

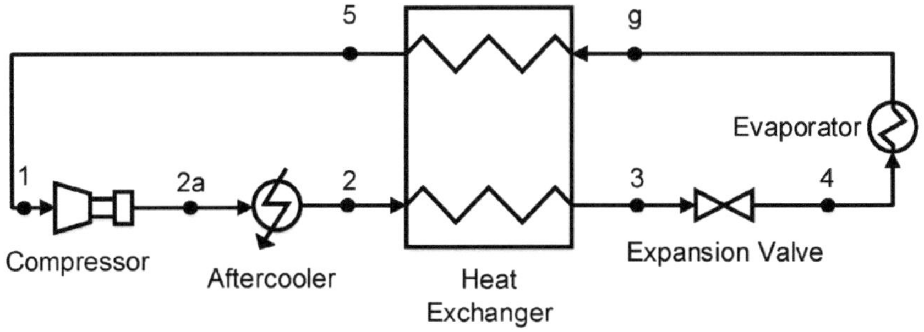

Figure 3.
Schematic diagram of joule-Thomson refrigeration (single MR) cycle.

Here, the refrigerant, which is in a vapor state at high pressure (2–3), is condensed through the heat exchange with a two-phase refrigerant (g–5) of low temperature and low pressure, and is transformed into a refrigerant of high pressure and low temperature. The high-pressure, low-temperature refrigerant that has passed through the intermediate exchanger passes through the expansion valve and transforms into a two-phase refrigerant of low temperature and low pressure because the pressure and temperature are decreased by the Joule-Thomson effect (3–4). Following this, the low-temperature, low-pressure refrigerant absorbs heat from the evaporator (4–g), passes through the intermediate heat exchanger, and is sucked into the compressor to complete the cycle.

The selected refrigerant type and composition proportions of the MR used in the single MR refrigerator vary depending on the evaporation temperature and operating conditions of the system. The evaporation temperature of the MR decreases as the proportion of low boiling point refrigerant increases, and the refrigeration capacity increases as the proportion of high boiling point refrigerant increases. Furthermore, as the two-phase sections of the selected refrigerants overlap, the time for reaching the target evaporation temperature decreases. The target evaporation temperature may not be reached if the selected composition proportions used in the MR are not appropriate, resulting in a stagnant temperature. Therefore, a specific method is required to select appropriate composition proportions of the MR. To select the composition proportions of an actual MR, it is essential to conduct experiments to validate various composition proportions selected theoretically.

2.2 Cascade mixed refrigerant (MR) refrigerator

A cascade MR refrigerator is a system that applies the Joule-Thomson cycle, in which the MR is applied to the two-stage cascade refrigeration system. With the active progress of the semiconductor market, the demand for refrigerators has increased, and related studies are an increasing trend. As mentioned, the cascade MR refrigerator includes an intermediate heat exchanger in the low-stage cycle of a two-stage refrigeration system. **Figure 4** shows the schematic diagram of the device.

The cascade MR refrigerator cycle is divided into high-stage and low-stage cycles, similar to a typical two-stage refrigeration cycle. First, the flow of the refrigerant in the high-stage cycle is introduced. The high-temperature, high-pressure vapor refrigerant discharged from the high-stage compressor becomes saturated or sub-cooled liquid as it passes through the condenser (1–2). The liquid refrigerant that passed through the expansion valve (2–3), then transforms into a two-phase low-temperature and low-pressure refrigerant. It flows into the cascade heat exchanger and exchanges

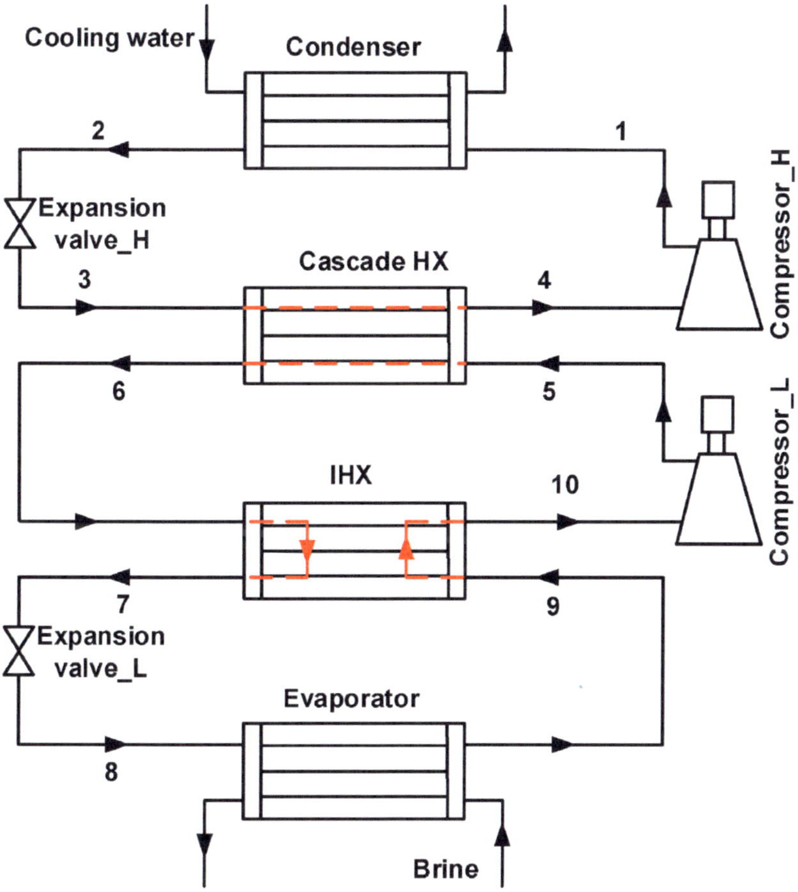

Figure 4.
Schematic diagram of cascade MR refrigeration cycle.

heat with the refrigerant flow of the low-stage cycle (3–4). In other words, from the perspective of the low-stage cycle, the cascade heat exchanger serves as the aftercooler of the previously introduced single MR. Next, the flow of the refrigerant in the low-stage cycle is discussed. The refrigerant discharged from the low-stage compressor in a state of high-temperature, high-pressure vapor is partially condensed (5–6), which is then condensed into a fully liquid refrigerant through the internal heat exchange in the intermediate heat exchanger (6–7) and passes through the expansion valve (7–8). The liquid refrigerant that has passed through is in a state of low temperature and exchanges heat with the brine in the evaporator, resulting in partial evaporation (8–9). Complete evaporation occurs as the intermediate heat exchanger absorbs the heat from the high-pressure refrigerant (9–10). The evaporated refrigerant is sucked by the compressor, causing the cycle to repeat itself (10–5).

Meanwhile, it is required to explain the concept of Joule-Thomson cooling capacity, which is an important concept in studying the refrigerant composition in the MR refrigeration cycle. **Figure 5** shows an example of the refrigeration cycle using a single refrigerant for clarity. The enthalpy difference occurring when expanding from a high pressure to a low pressure along the isotherm is referred to Joule-Thomson cooling capacity. **Figures 6** and 7 show the Joule-Thomson cooling capacity of various

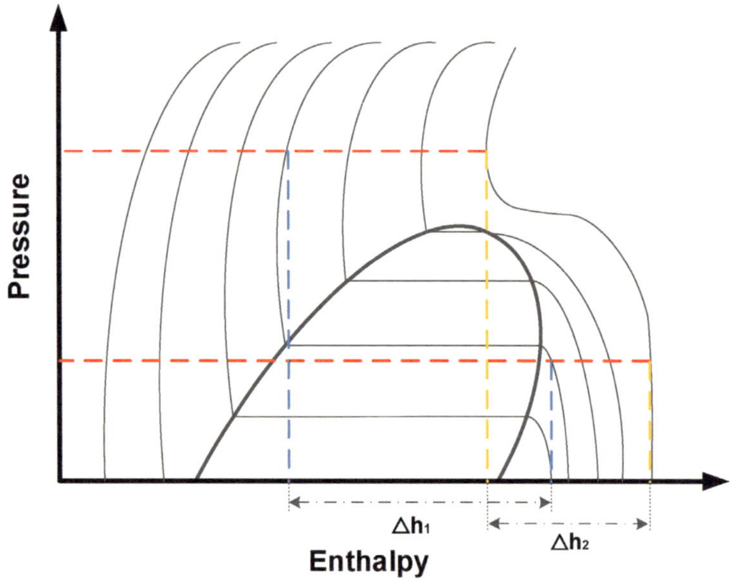

Figure 5.
P-h diagram of pure refrigerant for explaining joule-Thomson effect.

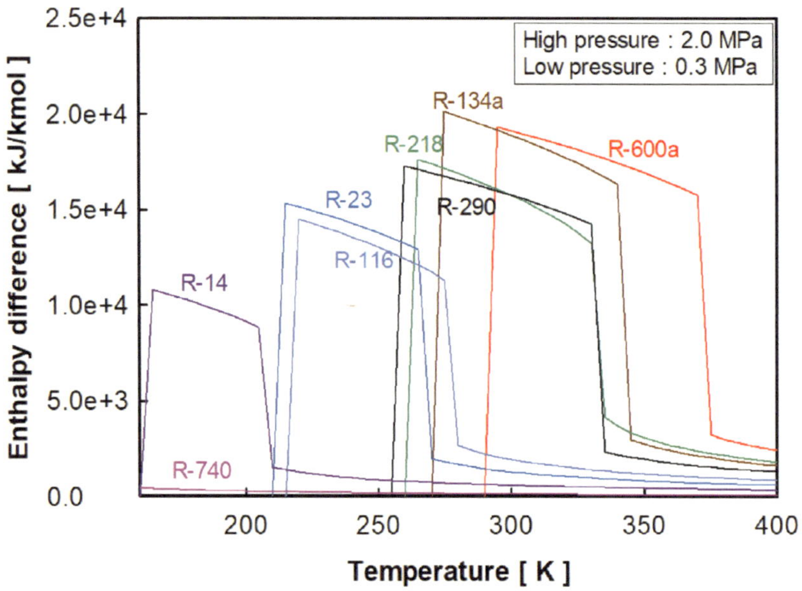

Figure 6.
Enthalpy difference according to temperature in kJ/kmol unit [13].

refrigerants used in the composition of MR at each temperature point. For the refrigerants used in the analysis, **Table 1** shows the normal boiling point, global warming potential (GWP), ozone depletion potential (ODP), and the refrigerant safety group through the American Society of Heating, Refrigerating and Air-Conditioning Engineers (ASHRAE) 2009 [14].

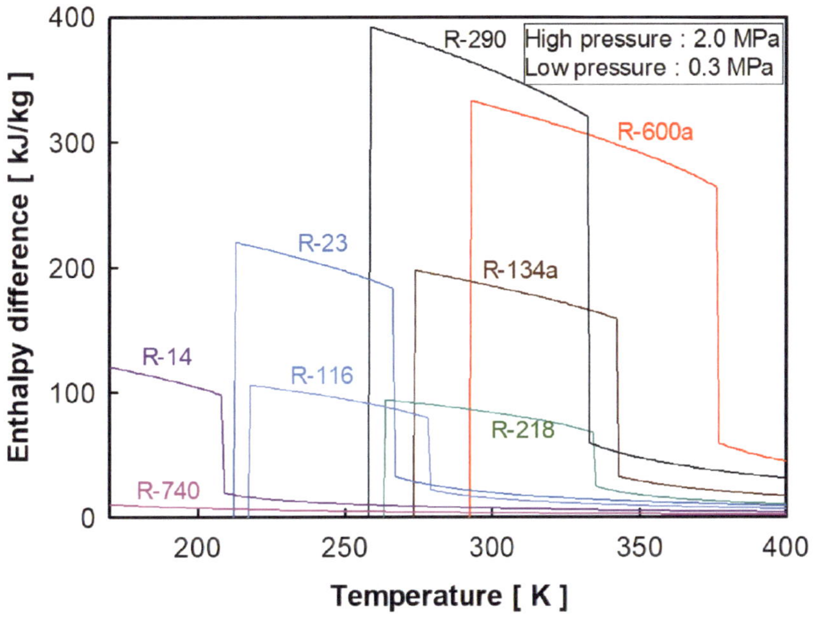

Figure 7.
Enthalpy difference according to temperature in kJ/kg unit [13].

Type	Refrigerant	Normal boiling point	GWP	ODP	Safety group	Molecular weight
HC	R-600a	−11.7	3	0	A3	58.122
HFC	R-134a	−26.3	1,430	0	A1	102.03
PFC	R-218	−36.7	8,830	0	A1	188.02
HC	R-290	−42.1	3.3	0	A3	44.096
PFC	R-116	−78.2	12,200	0	A1	138.01
HFC	R-23	−82.1	14,800	0	A1	70.014
PFC	R-14	−127.8	7,390	0	A1	88.010

Table 1.
Properties information of various refrigerants [14].

As shown in **Figure 6** [13], in general, a refrigerant with a low standard boiling point has a small enthalpy difference, and a refrigerant with a high standard boiling point has a large enthalpy difference. If the temperature to be reached is low, then a refrigerant with a low standard boiling point should be used. However, because the temperature must be reduced from room temperature to a low-temperature region, a refrigerant with a high boiling point should be mixed to utilize advantage of a large enthalpy difference of it. In other words, a refrigerant with a low boiling point and that with a high boiling point should be mixed appropriately to satisfy the target cooling capacity and temperature simultaneously.

Figure 7 shows the enthalpy difference per unit mass for the same refrigerants by converting the y-axis from the unit mole to the unit mass. The refrigerants with high enthalpy differences, such as i-Butane (R600a) and propane (R290) refrigerants, which are hydrocarbon (HC) refrigerants, appear prominently regardless of the standard boiling point. These HC refrigerants have a characteristic that the molecular weight is small compared with other refrigerants, and when it is represented by the unit mass, a larger number of moles is included. Therefore, HC refrigerants can have larger enthalpy differences.

3. Experimental system and methodology

This section describes the experimental system and method of the cascade MR refrigerator, which is one of several refrigeration systems commonly used to achieve the ultra-low temperature introduced in Section 2.

Figure 8 shows the schematics of the experimental system, which is divided into low and high stage cycles. The high-stage cycle comprises of a compressor, condenser, expansion valve, and cascade heat exchanger, while the low-stage cycle comprises of a compressor, expansion tank, pre-cooling heat exchanger, cascade heat exchanger, intermediate heat exchanger, expansion valve, and evaporator. The cooling water side consists of an isothermal bath, process chilled water (PCW) pump, and inverter, and the flow rate of the cooling water is adjusted by controlling the number of revolutions. The brine side consists of a brine tank, brine pump, inverter, and heater. The inverter also controls the flow rate of the brine, and the load is controlled using the heater in the brine tank.

Figure 9 explains the use of the pre-cooling heat exchanger, which was not mentioned in the aforementioned basic description of the cycle. The pre-cooling heat exchanger uses the cooling water going into the condenser of the high-stage cycle to primarily cool the high-temperature, high-pressure refrigerant discharged from the compressor of the low-stage cycle, and it handles a portion of the after-cooler capacity. A simulation analysis was performed to determine whether to use

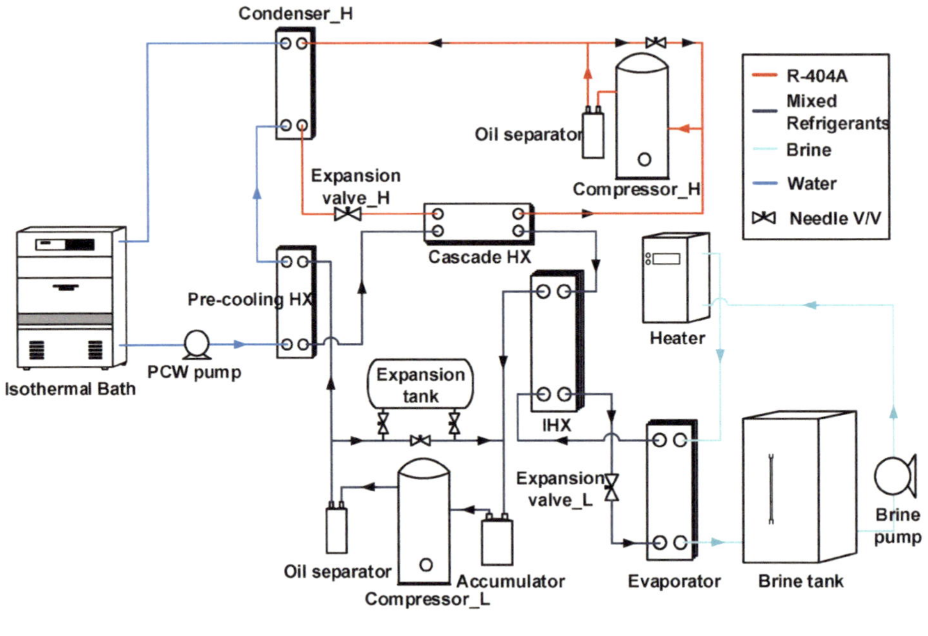

Figure 8.
Schematics of experimental system of cascade MR refrigerator.

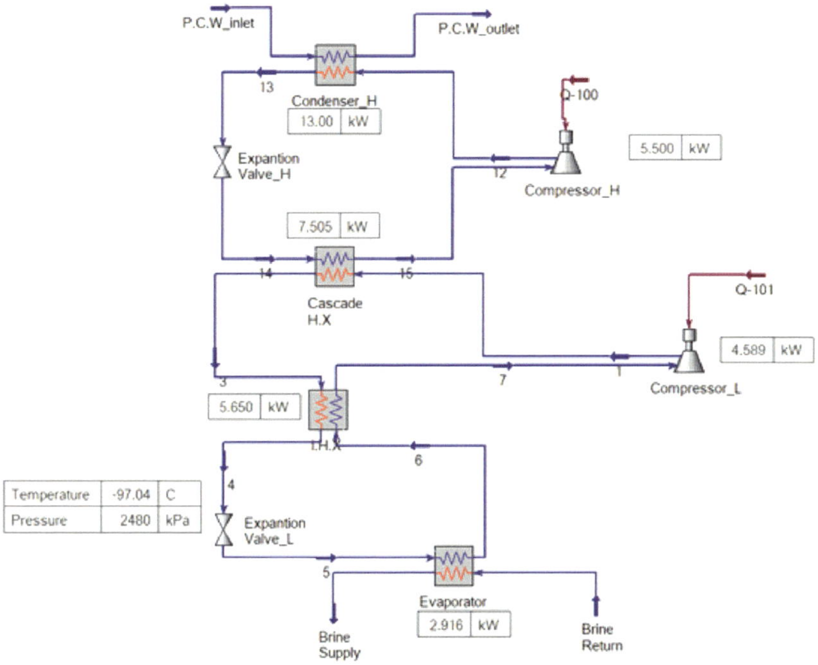

(a) without pre-cooling heat exchanger

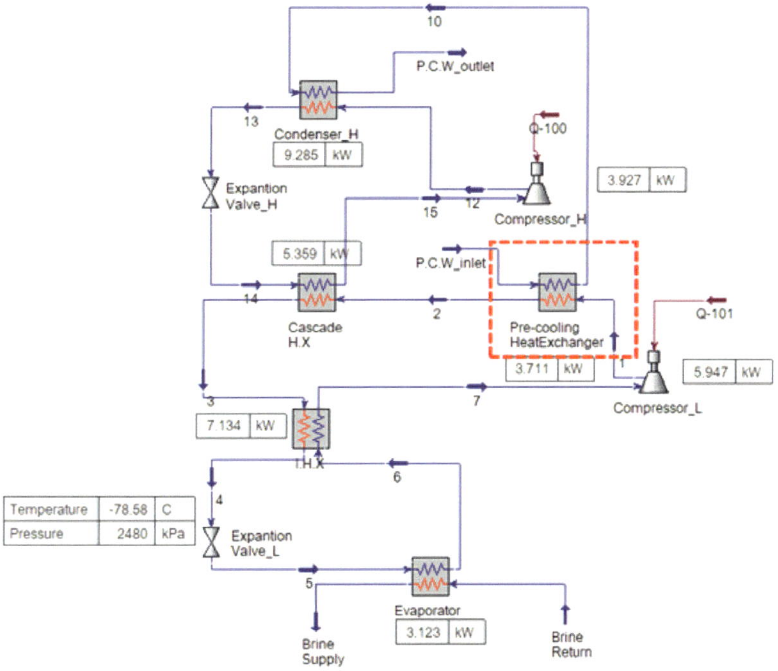

(b) with pre-cooling heat exchanger

Figure 9.
Comparison of cascade MR refrigerator with and without pre-cooling heat exchanger.

the pre-cooling heat exchanger. It should be noted that its use is accountable for a portion of the capacity handled by the aftercooler, and
simultaneously has an adverse effect on the temperature of the cooling water entering the condenser of the high-stage cycle. The results of the simulation analysis show that the application of the pre-cooling heat exchanger increases the coefficient of performance (COP) by approximately 10% from 0.289 to 0.316. The actual device produced and experimented based on this is shown in **Figure 10**.

The experiments were carried out under the conditions shown in **Table 2** to determine the performance characteristics according to the MR composition of the cascade MR Joule-Thomson refrigerator that uses the MR as the working fluid. Cooling water of 18°C flows through the pre-cooling heat exchanger at 22 L/min and is supplied to the high-stage condenser, and the brine entering the evaporator is supplied at a constant flow rate of 40 L/min. The compressor selected was a standard commercial compressor. The discharge pressure should not exceed 3 MPa, and the suction pressure should be maintained above 0.1 MPa, according to the operational constraints of the compressor.

Figure 10.
Experimental device of cascade MR refrigerator.

Parameter	Value	Unit
Cooling water inlet temperature	18	°C
Cooling water volume flow	22	L/min
Brine volume flow	40	L/min
Brine supply temperature	−100	°C
Compressors discharge pressure constraint	3.0	MPa
Compressors discharge temperature constraint	120	°C
Compressor_L suction pressure constraint	0.1	MPa

Table 2.
Summary of experimental conditions.

For reference, if the suction pressure is in a vacuum state, the oil supply in the compressor may not be smooth and moisture may enter. Therefore, the suction pressure of the compressor must be higher than vacuum pressure. To prevent the carbonization of the compressor oil, the discharge temperature is limited to be within 120°C.

The following method was used under the aforementioned experimental conditions. If the high-stage cycle reaches the target temperature, the compressor of the low-stage cycle is started. To prevent excessive pressure buildup, the valves installed before and after the expansion tank are opened during the process. The brine pump is operated as soon as the compressor starts, thereby exchanging the heat at the evaporator. Following this, the opening of the expansion valve is controlled in such a way that the suction pressure of the compressor does not drop to or below the atmospheric pressure, until the target temperature is reached. As the target temperature is approached, the temperature of the low-pressure side stream entering the intermediate heat exchanger decreases, and accordingly, the pressure of the high-pressure side stream decreases. To solve this problem, the valve of the pipe connected to the compressor suction tube and the expansion tank is opened to inject the refrigerant to maintain the pressure. If the target temperature

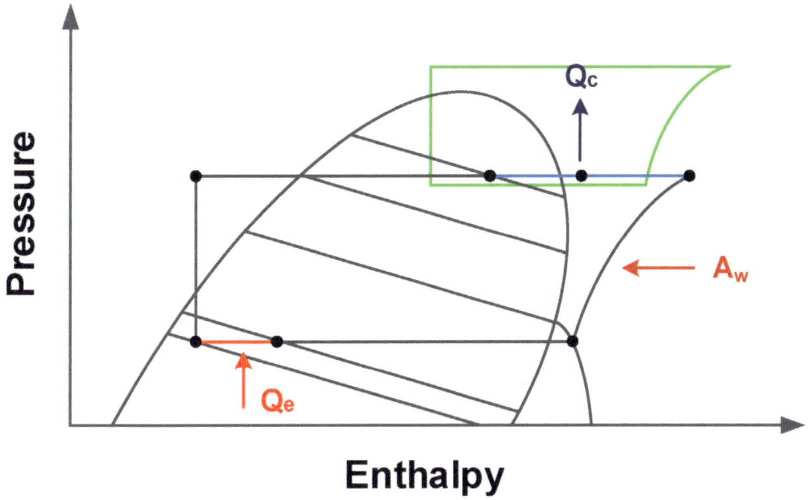

Figure 11.
P-h diagram of cascade MR refrigerator.

is reached, the load of the heater in the brine tank is gradually increased, and if the steady state is reached, the cooling capacity, including the load put on the heater and the heat generation pump, is measured.

Figure 11 shows the P-h diagram of the cascade MR refrigerator. The working fluid of the low-stage cycle consists of MR. When the flow rate of the cooling water entering the pre-cooling heat exchanger is sufficient, primary cooling proceeds up to the temperature of the cooling water, and the outlet temperature of the cascade heat exchanger is determined based on the evaporation temperature of the high-stage cycle. Here, when the low-stage cycle is viewed as a system, the sum of the cooling capacity at the evaporator and the work put into the compressor should be the same as the sum of the heat exhausted by the pre-cooling heat exchanger and the cascade heat exchanger. As the capacity of the pre-cooling heat exchanger and the aftercooler increases, the capacity of the evaporator increases. Because of its characteristics, the Joule-Thomson cycle using MR has a temperature gradient within the two-phase region, and the refrigerant temperature at the aftercooler outlet is limited to the temperature level of the cooling water. Therefore, in this study, the evaporator of the high-stage cycle, that is, the cascade heat exchanger, is used as the aftercooler of the low-stage cycle to increase the cooling capacity of the evaporator.

4. Experimental results and discussions

This section provides a brief overview of the data obtained using the experimental device of the cascade MR refrigerator introduced in Section 3. The specific composition proportions of the MRs below are not specified because they are protected by patent rights. In terms of changing the composition proportions, the total charging mass of MR was kept constant, and the mass proportions of the high boiling point refrigerants.

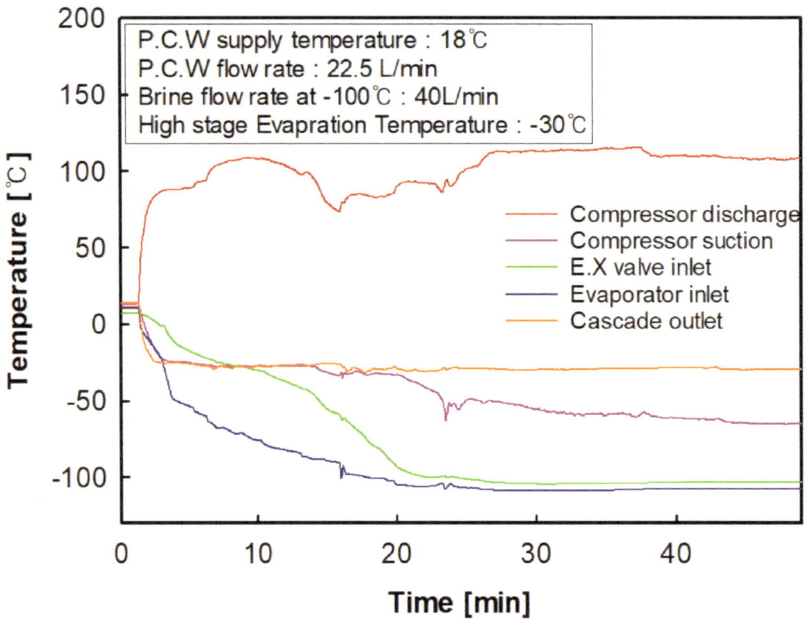

Figure 12.
Transient temperature of each measuring point (including R290).

R290 and R600a were adjusted. In other words, the mass of three different types of refrigerants was increased by the same proportion as the reduced charging amount of the high boiling point refrigerant, and the total charging amount of the MR remained constant [15].

4.1 Experiments with MR compositions of R290, R116, R23, and R14

Figure 12 shows the compressor discharge temperature and suction temperature, evaporator inlet temperature, and expansion valve inlet temperature according to the operating time in the refrigerant's initial composition proportion (R290, R116, R23, and R14) state. The device was operated by applying the MR composed of the above four refrigerants, and the opening of the expansion valve was controlled according to the rapid pressure change. Here, the temperature was decreased while leaving the suction valve of the expansion tank open, and as the target temperature was reached, the valve was closed, and the brine heater was operated.

The experiment was conducted while adjusting R290 in a mass fraction range of 5–50%, as shown by MR1–MR8, to examine the performance of the device based on the proportion change of the high boiling point refrigerant (R290) in the refrigerant charging amount of the same mass. The charging amount of R290 was the highest in MR1 and the lowest in MR8. **Figure 13** shows the compressor discharge and suction temperature, evaporator inlet temperature, and expansion valve inlet temperature based on the composition ratio of R290. When the mass ratio of R290 decreased, the compressor suction temperature increased from −67.8°C to −48.2°C. In both cases of MR7 and MR8, the expansion valve inlet temperature was clearly high when the mass fraction of the high boiling point refrigerant dropped below a certain level. However, the evaporator inlet temperature after going through the expansion valve was between −109.2 and − 106.7°C, showing no significant effect.

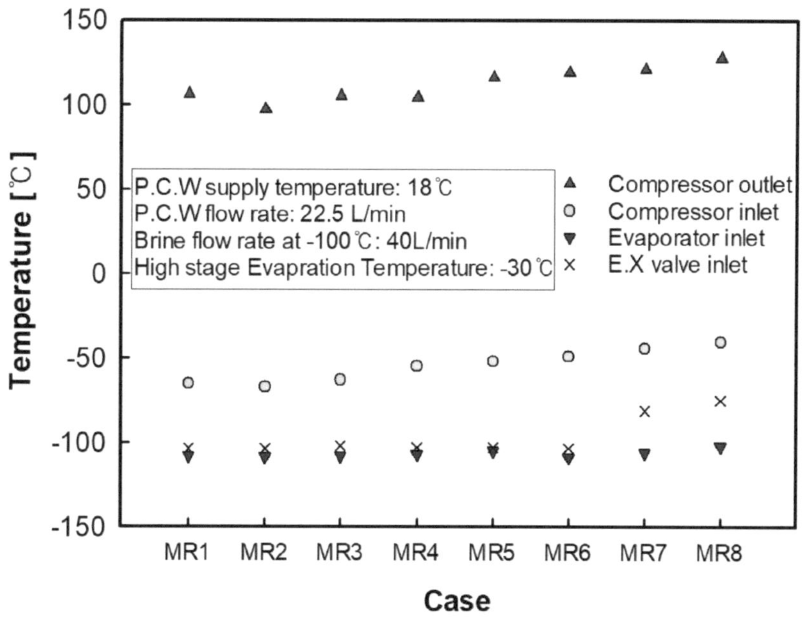

Figure 13.
Temperature comparison of various refrigerant composition cases (including R290).

Figure 14 shows a graph by comparing the changes in the cooling capacity and time when the composition of the high boiling refrigerant is changed. As the proportion of the high boiling point refrigerant increases, the cooling capacity shows a decreasing tendency. MR8 shows the highest cooling capacity at the 2.36 kW level, whereas MR1 shows the lowest value with 1.69 kW. However, there is a disadvantage in that the average pressure of the device increases because of the mid/low boiling point refrigerant that has a relatively higher pressure at room temperature as the proportion of the high boiling point refrigerant decreases. Furthermore, the cooling time was defined as the time when the brine supply temperature reaches −100°C, as measured through experiments. As the mass fraction of R290, a high boiling point refrigerant, increased, faster cooling time was achieved.

4.2 Experiments with MR compositions of R600a, R116, R23, and R14

Experiments based on the composition proportion of high boiling point refrigerant were carried out by replacing R290, the high boiling point refrigerant in the initial refrigerant, with R600a. In this case, the total charging amount in the mass of the refrigerant was the same.

Figure 15 shows the results of experiments performed by adjusting R600a to a mass fraction of 10–35%. Unlike the experimental results of the R290-applied MR, the compressor suction temperature and discharge temperature did not change significantly according to the composition change of the high boiling point refrigerant, in the results of R600a-applied MR experiments. Furthermore, the evaporator inlet temperature was maintained relatively constant between −109.2 and − 108.4°C.

Figure 16 shows the cooling capacity and cooling time according to the decrease of mass ratio of the high boiling point refrigerant. As the mass ratio of the high boiling point refrigerant in the composition of the MR decreased, the cooling capacity

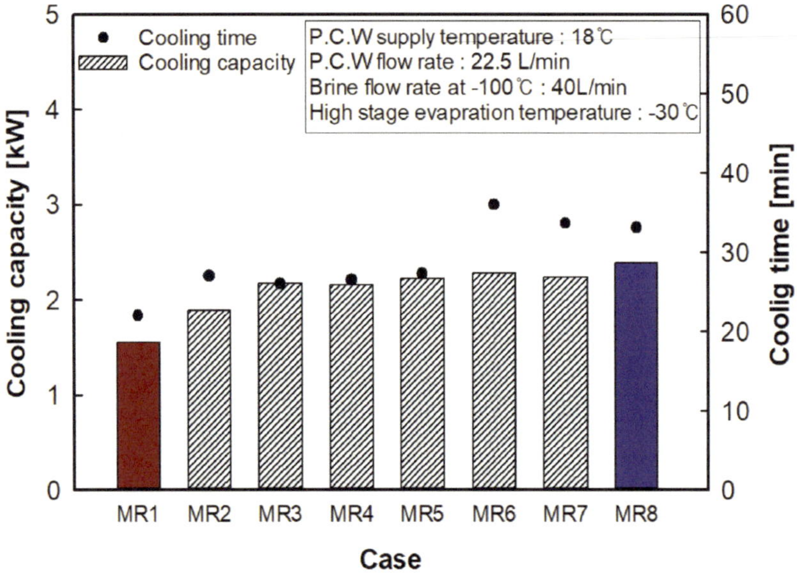

Figure 14.
Cooling capacity and time comparison of various refrigerant composition cases (including R290).

increased to 1.94–2.27 kW, and the cooling time was 37–46 min, indicating that the target temperature was reached at a slower rate compared to the MR using R290. It appears that a refrigerant with a boiling point between the high and low boiling points needs to be added to reach the cooling time faster.

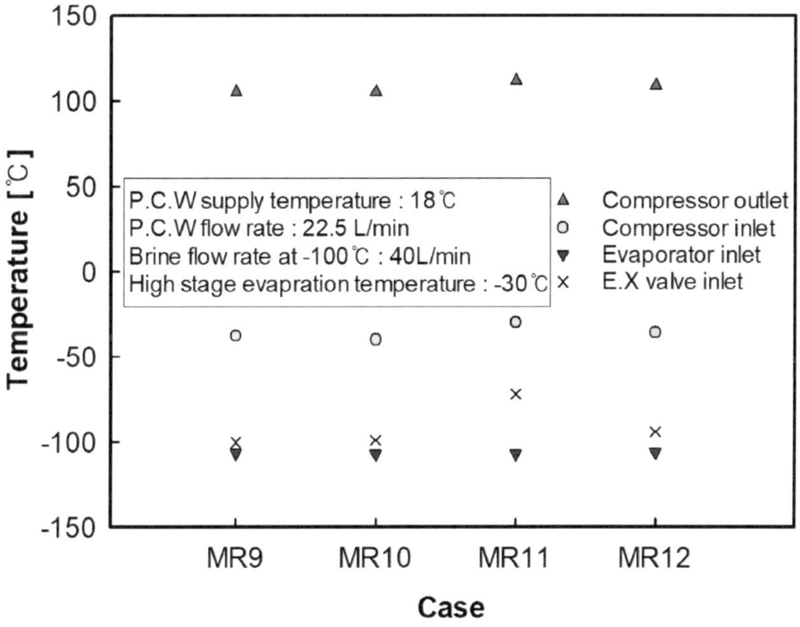

Figure 15.
Temperature comparison of various refrigerant composition cases (including R600a).

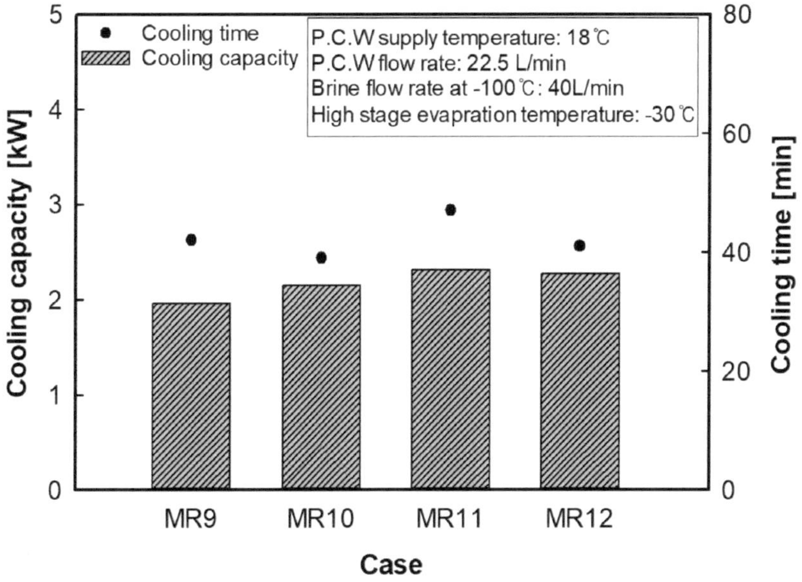

Figure 16.
Cooling capacity and time comparison of various refrigerant composition cases (including R600a).

5. Conclusions

Four different types of working fluids were selected for low-temperature production of below −100°C, which is required in the semiconductor manufacturing process, and experiments were carried out for R290 and R600a, which are HC refrigerants that show the largest enthalpy differences per unit mass among the high boiling point refrigerants of R290, R600a, R218, and R134a. The effects of changing the composition of the high boiling point refrigerant when the total charging amount is the same in both refrigerant groups was analyzed, and the results show that there is an advantage: as the mass ratio of the high boiling point refrigerant increases, the time for reaching the target temperature decreases, and the average pressure in the device is low. However, as the mass ratio of the high boiling point refrigerant decreases, the cooling capacity increases. The MR shows that the time for reaching the target temperature is short, and the cooling capacity is high when R290 is used as the high boiling point refrigerant among R290 and R600a.

Author details

Jung-In Yoon, Chang-Hyo Son, Sung-Hoon Seol* and Ji-Hoon Yoon
Department of Refrigeration and Air-Conditioning Engineering, College of Engineering, Pukyong National University, South Korea

*Address all correspondence to: seolsh@pknu.ac.kr

References

[1] K. Frost, I. Hua, Quantifying spatiotemporal impacts of the interaction of water scarcity and water use by the global semiconductor manufacturing industry, Water Resour. Ind. 22 (2019) 100115. https://doi.org/10.1016/j.wri.2019.100115.

[2] C.F. Chien, Y.J. Chen, Y.T. Han, Y.C. Wu, Industry 3.5 for optimizing chiller configuration for energy saving and an empirical study for semiconductor manufacturing, Resour. Conserv. Recycl. 168 (2021) 105247. https://doi.org/10.1016/j.resconrec.2020.105247.

[3] M. Newport, Semiconductor Lithography Overview, (n.d.). https://www.newport.com.cn/n/photolithography-overview.

[4] S.K.S. Fan, C.Y. Hsu, C.H. Jen, K.L. Chen, L.T. Juan, Defective wafer detection using a denoising autoencoder for semiconductor manufacturing processes, Adv. Eng. Informatics. 46 (2020) 101166. https://doi.org/10.1016/j.aei.2020.101166.

[5] K.H. Baek, T.F. Edgar, K. Song, G. Choi, H.K. Cho, C. Han, An effective procedure for sensor variable selection and utilization in plasma etching for semiconductor manufacturing, Comput. Chem. Eng. 61 (2014) 20-29. https://doi.org/10.1016/j.compchemeng.2013.09.016.

[6] H. Liang, Y. Chen, X. Xia, C. Zhang, R. Shen, Y. Liu, Y. Luo, G. Du, A preliminary study of SF6 based inductively coupled plasma etching techniques for beta gallium trioxide thin film, Mater. Sci. Semicond. Process. 39 (2015) 582-586. https://doi.org/10.1016/j.mssp.2015.05.065.

[7] H. Abe, M. Yoneda, N. Fujiwara, Developments of plasma etching technology for fabricating semiconductor devices, Jpn. J. Appl. Phys. 47 (2008) 1435-1455. https://doi.org/10.1143/JJAP.47.1435.

[8] L.W. Foo, Radimin, M. Teo, C. Lee, FEA thermal investigation of wafer thinning by plasma etching, 2006 Int. Conf. Electron. Mater. Packag. EMAP. (2006) 815-819. https://doi.org/10.1109/EMAP.2006.4430570.

[9] S.B. Radovanov, P. Corey, G. Angel, D. Brown, Wafer floating potential for a high current serial ion implantation system, Proc. Int. Conf. Ion Implant. Technol. 1 (1999) 482-485. https://doi.org/10.1109/iit.1999.812157.

[10] K. Yasui, M. Osaki, A. Miyamoto, H. Namai, Three-dimensional structure recognition of circuit patterns on semiconductor devices using multiple SEM images detected in different electron scattering angles, Microelectron. Reliab. 108 (2020) 113628. https://doi.org/10.1016/j.microrel.2020.113628.

[11] V. Gadhiraju, Cryogenic Mixed Refrigerant Processes, Springer New York, New York, NY, 2008. https://doi.org/10.1007/978-0-387-78514-1.

[12] O. Podtcherniaev, Performance of throttle-cycle coolers operating with mixed refrigerants designed for industrial applications in a temperature range 110 to 190 K, in: AIP Conf. Proc., AIP, 2002: pp. 863-872. https://doi.org/10.1063/1.1472105.

[13] K.-S. Lee, J.-I. Yoon, C.-H. Son, J.-H. Lee, C.-G. Moon, W.-J. Yoo, B.-C. Lee, Performance Characteristics of a Joule–Thomson Refrigeration System with Mixed Refrigerant Composition, Heat Transf. Eng. (2020) 1-10. https://doi.org/10.1080/01457632.2020.1776995.

[14] ASHRAE, 2009 ASHRAE Handbook-Fundamentals, American Society of Heating, Refrigerating and Air-conditioning Engineer, 2009.

[15] Ji-Hoon Yoon, Performance characteristics of cascade Joule-Thomson Refrigeration cycle with mixed refrigerants composition, Pukyong National University, n.d.

Section 2

Nonconventional Energy Recovery

Chapter 4

Assessment of Solar Energy Potential Limits within Solids on Heating-Melting Interval

Petrica Vizureanu, Madalina-Simona Baltatu, Andrei-Victor Sandu, Dragos-Cristian Achitei, Dumitru-Doru Burduhos-Nergis and Manuela-Cristina Perju

Abstract

The solar furnace works by using the electric energy produced by a photovoltaic system, which converts solar energy, solar radiation, into electric energy. The performances of the solar furnace used in various applications from industry are influenced by various factors. One of these factors imposes the acquisitions of certain large densities of the radiant power, and it requires a geometric form of the concentrator. The research is based on the behavior of some metallic alloys at elevated temperatures, for purifying some materials and for the achievement of some chemical synthesis. An important technological condition is a temperature which is achieved by concentrating solar radiation. This temperature is necessary to produce metallic material in the crucible, without other complementary energy for the thermal process. Steel or aluminum production requires very high quantities of thermal energy. Usually, this energy is given by electric power, natural gases, or conventional fuels. The solar furnace uses the energy given by the sun. For the manufacturing of the electrothermal furnaces, a series of specific materials are used, which are necessary for the obtaining of the furnace chamber, for the heating elements, as well as for the measurement systems of the temperature.

Keywords: solar furnaces, climatic conditions, concentrator, applications, sun's rays

1. Introduction

Many of the photovoltaic systems function independent of the wiring system (off-grid). These systems are made of solar panels matrixes, control systems, storage systems, and DC or AC consumption devices. The scheme of a PV system is shown in **Figure 1**.

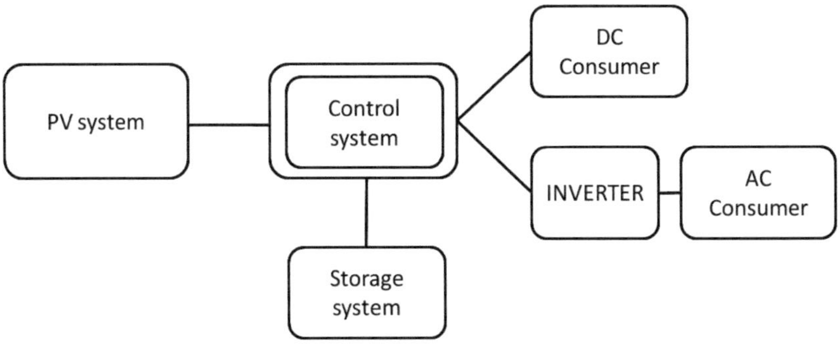

Figure 1.
Scheme of an off-grid PV system.

Panel matrixes are made of modules that consist of serially or parallelly connected panels depending on requirements. The panels are generally made of 36 serially connected cells (the number of cells commonly used are 32, 36, 48, 60, 72, and 96). For example, a solar panel comprising 32 cells typically can produce 14.72 volts output. Specific batteries are used for PV systems for storage and to stabilize tension. Control systems have charging regulators and converters/inverters which can be either DC-DC or DC-AC and blocking diodes. Control systems assure the interface between all the components of the PV system for protecting and controlling the system [1–6].

A solar furnace will use solar energy in the processes for technological purposes. This installation belongs to the thermal installations category. The advantages of thermal installations based on solar energy are as follows:

- In terms of conversion system, a powerful thermal source can be concentrated into a tight space with a direct consequence on getting some high temperatures (over 2.000°C);

- As in any heating installation based on electric power, installations based on solar cells also have the possibility of accurate and rapid temperature adjustment;

- Electric system used for the conversion of solar power into thermal power allows simple automation with high reliability;

- Transfer from classical energetic sources (solid or liquid fuels or electric power achieved through their burning) to alternative sources can be the future alternative required by the endearment of classical sources and their powerful pollution.

As a result of these electrothermal installations advantages, installations that rely on solar energy sources can be used in different industrial processes [7–17] such as:

- Fabrication technologies of ferrous or nonferrous materials;

- Metals heating for thermal processing;

- Manufacturing of various abrasive materials;

- Welding of metallic or nonmetallic alloys;

- Glass and ceramics processing;
- Drying, preparation, and sterilization of food products.

The documentary research visits made at Instituto de Soldadura e Qualidade, Lisabona, and at Ecole Européenne d'Ingénieurs en Génie des Matériaux, Nancy, opened new collaboration perspectives with renowned teams from abroad for achieving new results due to their experience in the domain. Another documentary research visit is made at the Dipartimento di Ingegneria Aerospaziale e Meccanica, Secunda Universitadegli Napoli, Italia.

To achieve the necessary elevated temperatures, an important condition is to use quality materials. The best option is the use of electric resistor furnaces.

2. Materials and methods

2.1 Documentary study regarding solar furnaces and the importance of climatic conditions

In order to design a PV system, we start from the consuming requisite power (load resistance) which in this case is an electric furnace with a capacity of 1 liter and a 0.55-kWh power, a lab furnace, the working time and climatic conditions of the zone where the system is installed (**Figure 2**).

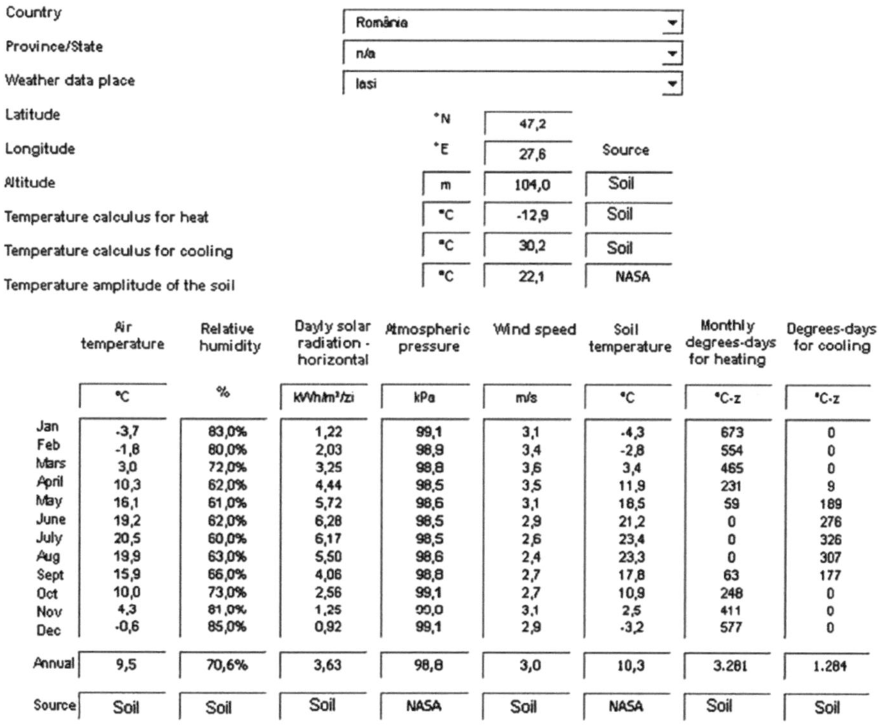

Country	România		
Province/State	n/a		
Weather data place	Iasi		
Latitude	°N	47,2	
Longitude	°E	27,6	Source
Altitude	m	104,0	Soil
Temperature calculus for heat	°C	-12,9	Soil
Temperature calculus for cooling	°C	30,2	Soil
Temperature amplitude of the soil	°C	22,1	NASA

	Air temperature	Relative humidity	Dayly solar radiation - horizontal	Atmospheric pressure	Wind speed	Soil temperature	Monthly degrees-days for heating	Degrees-days for cooling
	°C	%	kWh/m²/zi	kPa	m/s	°C	°C-z	°C-z
Jan	-3,7	83,0%	1,22	99,1	3,1	-4,3	673	0
Feb	-1,8	80,0%	2,03	98,9	3,4	-2,8	554	0
Mars	3,0	72,0%	3,25	98,8	3,6	3,4	465	0
April	10,3	62,0%	4,44	98,5	3,5	11,9	231	9
May	16,1	61,0%	5,72	98,6	3,1	18,5	59	189
June	19,2	62,0%	6,28	98,5	2,9	21,2	0	276
July	20,5	60,0%	6,17	98,5	2,6	23,4	0	326
Aug	19,9	63,0%	5,50	98,6	2,4	23,3	0	307
Sept	15,9	66,0%	4,06	98,8	2,7	17,8	63	177
Oct	10,0	73,0%	2,56	99,1	2,7	10,9	248	0
Nov	4,3	81,0%	1,25	99,0	3,1	2,5	411	0
Dec	-0,6	85,0%	0,92	99,1	2,9	-3,2	577	0
Annual	9,5	70,6%	3,63	98,8	3,0	10,3	3.281	1.284
Source	Soil	Soil	Soil	NASA	Soil	NASA	Soil	Soil

Figure 2.
The necessary climatic conditions.

PV electricity generation for: Nominal power=1.0 kW, System losses=24.0%				
	Inclin.=37 deg., Orient.=0 deg.		2-axis tracking system	
Month	Production per month (kWh)	Production per day (kWh)	Production per month (kWh)	Production per day (kWh)
Jan	44	1.4	54	1.7
Feb	61	2.2	77	2.7
Mar	94	3.0	120	3.9
Apr	99	3.3	131	4.4
May	119	3.8	169	5.4
Jun	115	3.8	167	5.6
Jul	122	3.9	174	5.6
Aug	123	4.0	171	5.5
Sep	103	3.4	135	4.5
Oct	86	2.8	109	3.5
Nov	48	1.6	58	1.9
Dec	37	1.2	45	1.5
Yearly average	**88**	**2.9**	**117**	**3.9**
Total yearly production (kWh)		**1050**		**1410**

Figure 3.
Comparison between a fixed system with a 37° optimum angle and a tracking system.

As observed from the column which gives average horizontal solar radiation per day, we have the sufficient radiation quantity in order to produce consuming requisite power of 0.55 kWh. By adjusting the system to an optimum bend degree, a significant radiation increase can be obtained. If a tracking system is used, a higher quantity with almost 40% can be obtained, as shown in **Figure 3**, where considered system losses are also considered.

In order for a PV system to obtain a standard power of 1 kW, eight-module Mitsubishi PV-MF 130 EA2LF type, polycrystalline silicon, 8 x 130Wp = 1.04 kW should be used.

For this system, we use 3 batteries of 130Ah, 12 V with 2 days functioning reserve even without the sun (**Figure 4** and 5).

The calibration of the storage system is made based on the use period and the days when the system provides power from the batteries. The charging regulator for batteries has a basic function, that is, the charging stops when the batteries are completely charged.

There are some other functions such as consuming disconnection when the tension is small and temperature compensation [3–5].

Blockage diodes are used to avoid batteries discharge, on cells, during the night period or during cloudy days. For some systems, maximum power point trackers (MPPT) can be used.

The purpose of this device is to maintain the operating tension of the system to a maximum tension which is independent of the load resistance changes.

2.2 The determination of the geometric arrangement and the components of a solar furnace used for processing metallic and nonmetallic materials

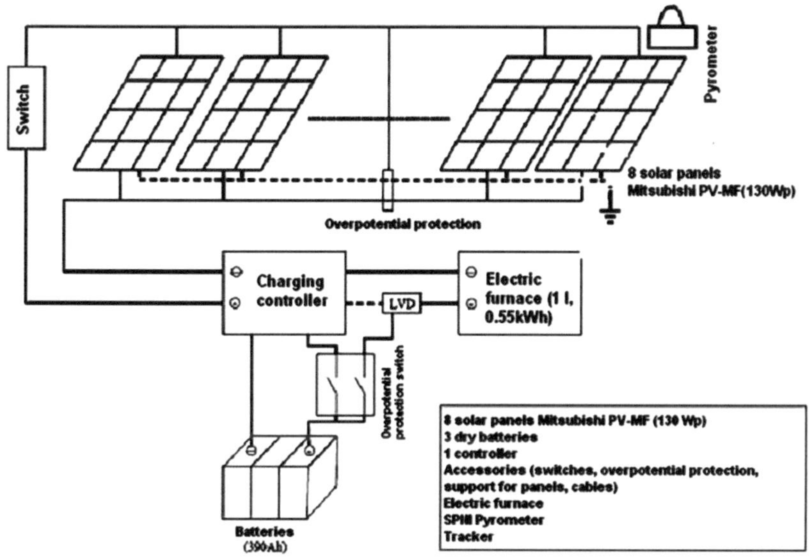

Figure 4.
Scheme of an electric system with PV for direct current.

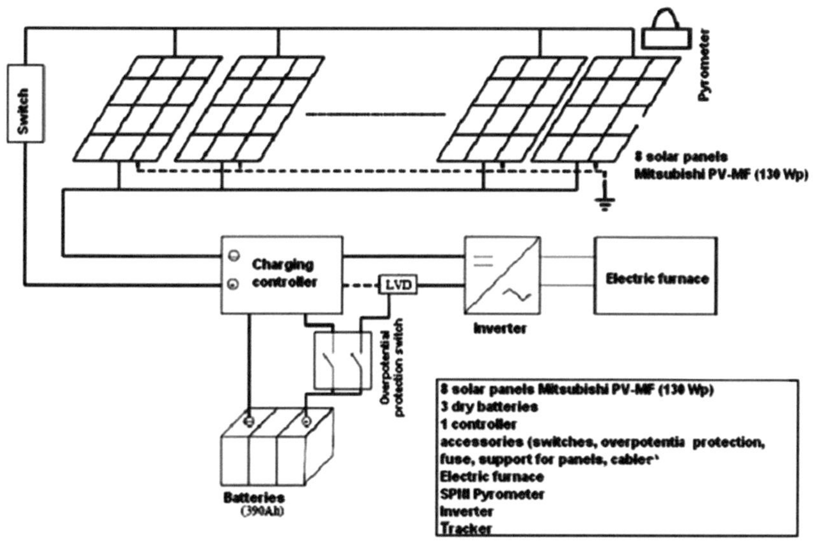

Figure 5.
Scheme of an electric system with PV for alternating current.

2.3 The analysis of heating process within solar furnaces

2.3.1 Mathematical model of the heating regime within the working chamber of a solar furnace

The factors that influence the performances of the solar furnace used in industrial applications that impose the procurement of some large densities of the radiant power and the need of a geometric perfection of the concentrator are classified in three categories as follows:

- The first type is pertinent to the basic geometry of the parabolic concentrator defined by focal distance, f, and its opening, D. Once these were chosen, it results in the dimensions of the solar image, focusing factor and the ideal maximum values of temperature and radiant power density into the image, regardless of the concentrator's construction and its installation place.

- The second type consists of the factors that reduce the performance of the solar furnace due to its construction and installation place. These factors are the transmission of the energetic factor of the atmosphere, the directional reflection energetic factor of the mirrors, and the index of geometric perfection of the parabolic concentrator.

- The third type includes the factors according to the receiver's properties: its adsorption and emission energetic factors and heat losses that take place through conduction and convection.

The available potential power P_f in sun's image from the focal plane of a solar furnace is given by the relation:

$$P_f = \pi \bullet R_d \bullet D_a \bullet E_0 \bullet f^2 \bullet \sin^2 \theta_{max} \tag{1}$$

where:

R_d – directional reflection energetic factor of the parabolic mirror (including heliostats, if they exist);

D_a – transmission energetic factor of the atmosphere where the furnace is installed;

$E_0 = 1353\ \mathrm{W/m^2}$ – (constant);

f – focal distance;

θ_{max} – the opening angle of the parabolic concentrator.

If a solid corp is disposed in the solar image, the available fraction of potential power effectively absorbed by the corps would be determined by the absorption factor and the form of this corps-receiver. As such, the maximum temperature that could be obtained into a solar furnace depends on the properties of the receiver disposed of in the focal zone of the furnace:

$$T_{max} = T_s \bullet (R_d \bullet D_a)^{1/4} \bullet (\sin \theta_{max})^{1/2} \tag{2}$$

where $T_s = 5800\ K$ – temperature at sun's surface.

2.3.2 The determination of the technological parameters involved in the heating process of a solar furnace

The most important research is made on the behavior of metals and refractory materials at elevated temperatures, for purifying nonmetallic materials and for the achievement of some thermochemical synthesis.

One of the technological parameters is the temperature that is obtained by focusing on the sun's rays. The furnace uses the temperature in order to melt the metallic material in the crucible, without other complementary energy for the thermal process (**Figure 6**) [5, 18].

Steel or aluminum production needs very high quantities of energy. This is usually given by electric power, natural gases, or conventional fuels. A solar furnace uses the energy given by solar radiation.

We can see in the image how the sun's rays can be focused toward the crucible where the ore is. This is heated to a very high temperature until it melts.

Pollution is basically inexistent because solar energy is a pure form of energy. The melting materials with very high melting points are one of the main applications of solar furnaces.

The material melting on a portion whose area is approximately equal to the area of the sun's image can take place in case the exterior of solid material is exposed to very intense radiation from the focal zone of a solar furnace.

As heat enters the solid, the melted material quantity increases and forms a liquid cavity. Through such a process, it is possible to melt the material in the crucible; this happens because of the existence of a high-temperature gradient between melted material and the crucible's exterior.

In regular furnaces, the crucible is warmed from the exterior, and it has a high temperature continuously than the melted raw material. As a result, the crucible in such furnaces must be made of a material that is more refractory than the substance to be melted, as well as chemically inert toward the melted material.

As the melting point of the examined material rises above 2000°C, the difficulty of achieving these two conditions increases, as there are fewer options to avoid chemical reactions.

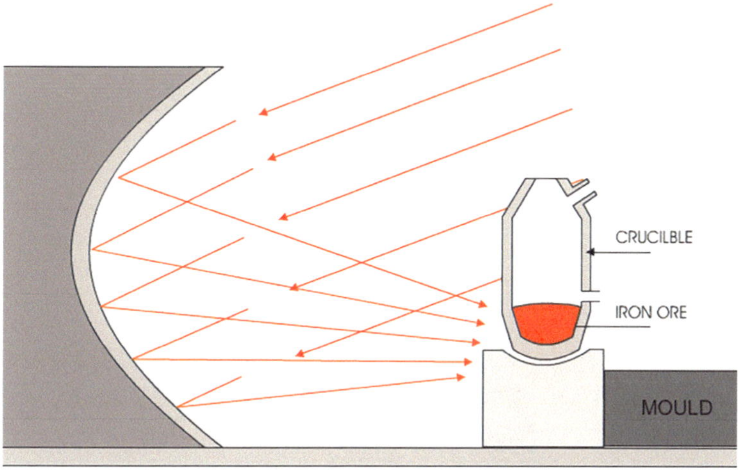

Figure 6.
The schematic representation of the installation that is used to melt a metallic material using solar energy.

Solar furnaces overcome these significant constraints of conventional furnaces when melting materials have high refractivity. As a result, melting can take place in furnaces with a horizontal axis.

The furnace is rotated around its horizontal axis and has an inner diameter several times greater than the diameter of the solar image. When the rotation speed is modest, the melted material stays in the lower part of the furnace, and the turning aids in heat distribution uniformity.

The melted material is centrifuged, generating a cavity that prevents it from flowing out of the furnace at higher rotation rates. The furnace's external walls, which are typically composed of steel, can be water-cooled to maintain (if necessary) a high-temperature gradient through the walls.

When melting into a specific protective environment, a suitable gas current is passed, as shown in **Figure 7**. Quartz, zirconium dioxide, corundum, ceramic oxides, and materials like carbides, nitrides, and boron are among the materials that can be examined. Conventional melting processes have many drawbacks for these materials.

It is also possible to investigate the feasibility of employing solar furnaces for steel melting. Technically, the crucible can be readily made by inserting a refractory powder into the furnace's cavity and sintering or even melting it through the furnace's top that is exposed to solar radiation. After that, scrap iron is inserted, melted, and then molted in forms if necessary [6, 7].

The performance of such a solar furnace does not need to be exceptional because there is sufficient temperature of 2000–2500°C.

Other metals that are more expensive than steel, such as titanium, zirconium, and molybdenum, are expected to generate increased attention in the future.

In this instance, an inert protective atmosphere must be ensured, and the challenges and costs associated with this must be considered.

Impurities evaporation, zonal melting, fractioned crystallization, the separation of zirconium oxide from zirconium (zirconium silicate), and material investigation under thermal shock conditions are some of the other applications of sun furnaces.

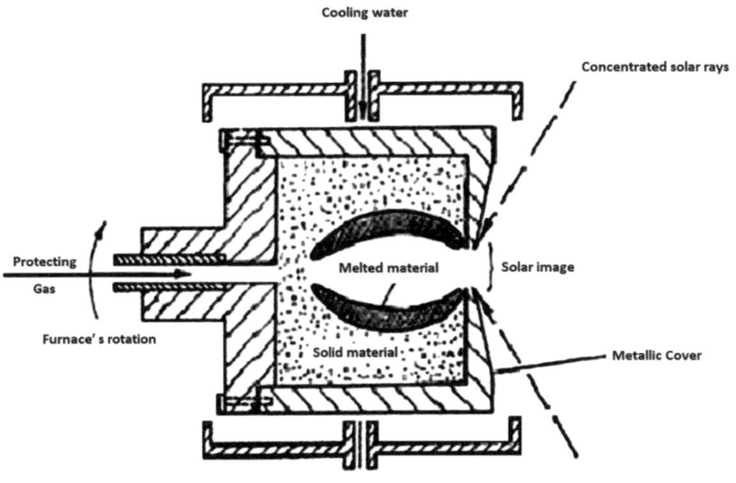

Figure 7.
Details of melting installation of metallic alloys using solar energy.

2.4 The analysis of melting/burning/purifying process within solar furnaces

These objectives consist in developing some mathematical formalism, which allows a better capitalization of the advantages given by the evolution of melting/burning/purifying within solar furnaces [14, 16].

Working out of this mathematical formalism was very useful for the documentary stage from Universidad de Las Palmas de Gran Canaria, Spain, as well as for the discussions on this theme with Prof. Agustin Santana Lopez.

2.4.1 Mathematical model of the melting/burning/purifying process within the working chamber of a solar furnace

Thermal efficiency η_t of an electrothermal installation based on solar energy is given by the ratio:

$$\eta_t = \frac{Q_u}{Q_u + Q_p + Q_a} \quad (3)$$

- Q_u is the quantity of **absorbed heat** necessary for heating the material;
- Q_p represents **heat loses** due to heating installation;
- Q_a is the heat quantity necessary for heating the auxiliary components of the installation;

The furnace will be in these working temperature classes that are considered as classification criteria in heating technology:

- Low-temperature furnace (between 600 and 700°C);
- High-temperature furnaces (until 1600°C).

In order to define the calculus model of the furnace based on solar energy, it is necessary to define the following **input data**:

- Material that will be heated with all its thermal and electric data;
- Charge shape and dimensions;
- Technological regime that consists in:
  - Heating time until reaching solidus temperature;
  - Heating temperature;
  - Overheating time for generalizing liquid state in the entire mass of the charge;
  - Overheating temperature;
  - Holding time at casting temperature necessary for eventual alloying in the liquid state;

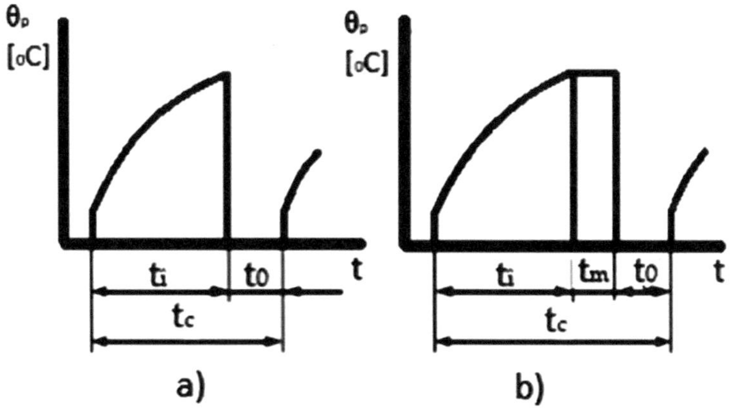

Figure 8.
Diagrams of possible functioning of the solar furnace: a) melting without holding a constant temperature; b) melting by holding at constant temperature (t_c – A complete cycle; t_i – Heating time; t_m – Holding time at constant temperature; t_o – Loading – Unloading time).

- Special technological conditions (protection atmosphere, vacuum, etc.);
- Furnace efficiency.

Heat consumption is calculated from functioning diagram of the furnace, that is, in fact, the variation diagram of load temperature–time function θ_p; taking into consideration the fact that this is a furnace with intermittent functioning, the diagrams presented in **Figure 8** are the possible ones.

2.4.2 Determination of technological parameters implied in melting/burning/purifying process within a solar furnace

a. **Temperature control**

Within the furnace (**Figure 9a**), the thermometric transducer T is installed that transmits information of the temperature to the AB adjusting block.

In comparator C, a tension proportional to the desired value of the temperature θ_d, determined on the basis of the program imposed by technological process and controlled by the block of desired values BVD, is compared with a tension proportional to the real value of the temperature within furnace θ_r.

If $\theta_r < \theta_d$, on–off regulator RBP is transmitting the closing command to adjusting block (connection switch to the power supply), and the furnace is absorbing power P. If $\theta_r > \theta_d$, cutting-out command of the switch is transmitted.

The adjustment made with a dead zone $\Delta\theta$ is given by the regulator's characteristic (**Figure 9b**).

b. **Orientation of photovoltaic system**

Adjusting the system to an optimum angle of inclination, a significant increase of radiation can be achieved, which can be used, instead of positioning the panels on horizontal or at a random angle in general, to place latitude.

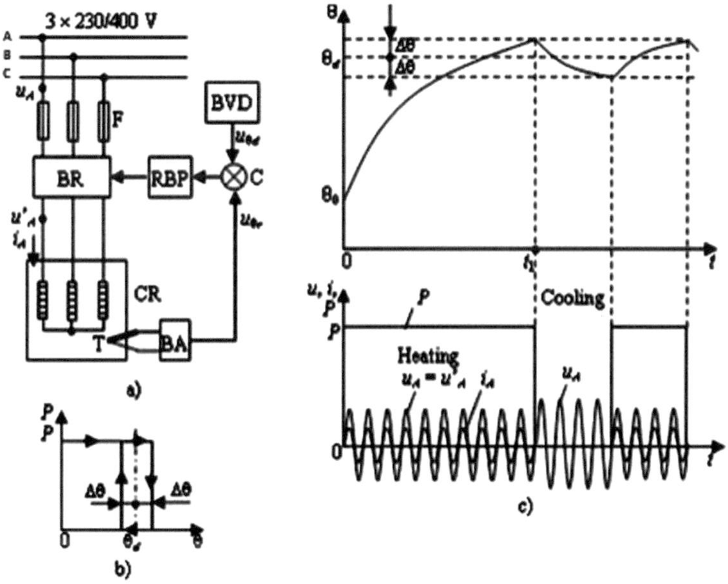

Figure 9.
On–off adjustment of furnace temperature.

If a tracking system is used, an increased quantity with approximately 40% will be achieved, as shown in **Figure 10**, where system losses are also considered.

The use of sensors for orientation can lead to delicate situations in case of sun-clouds alternations, if the system is not properly calibrated and has high energy consumptions.

Taking into account of these considerations, the variant that uses a mathematical algorithm is chosen for solar panel positioning. The orientation makes after the two directions, namely E-V and S-N.

PV electricity generation for: Nominal power=1.0 kW, System losses=24.0%				
	Inclin.=37 deg., Orient.=0 deg.		2-axis tracking system	
Month	Production per month (kWh)	Production per day (kWh)	Production per month (kWh)	Production per day (kWh)
Jan	44	1.4	54	1.7
Feb	61	2.2	77	2.7
Mar	94	3.0	120	3.9
Apr	99	3.3	131	4.4
May	119	3.8	169	5.4
Jun	115	3.8	167	5.6
Jul	122	3.9	174	5.6
Aug	123	4.0	171	5.5
Sep	103	3.4	135	4.5
Oct	86	2.8	109	3.5
Nov	48	1.6	58	1.9
Dec	37	1.2	45	1.5
Yearly average	88	2.9	117	3.9
Total yearly production (kWh)		1050		1410

Figure 10.
Comparison between a fixed system at an optimum angle of 37°and tracking system.

2.5 Characterization of metallic materials heating process within solar furnaces

2.5.1 Materials heating within a furnace

Equation of energetic balance for the furnace can be written as:

$$dQ_2 = dQ_u + dQ_a + dQ_{pd} + dQ_z \tag{4}$$

where:

dQ_2 is the elementary heat quantity transmitted toward furnace interior by the heating element:

$$dQ_2 = P_2 \cdot dt = \alpha \cdot A_1 \cdot (\theta - \theta_0) \cdot dt \tag{5}$$

dQ_u is the elementary heat quantity that leads to the heating of the useful material within the furnace (absorbed heat):

$$dQ_u = c_u \cdot m_u \cdot d\theta \tag{6}$$

dQ_a is theheat quantity that leads to heating the attached pieces (stands, supports, etc):

$$dQ_a = c_a \cdot m_a \cdot d\theta \tag{7}$$

dQ_{pd} is the elementary thermal losses through furnace walls, opening, leakiness, etc.;

dQ_z is the elementary heat quantity that gathers in furnaces walls:

$$dQ_u = c_z \cdot m_z \cdot d\theta \tag{8}$$

where:

P_2 – thermal energy (thermal flow) transmitted by photovoltaic systems;

α – heat exchange superficial coefficient;

A_l – area of the total lateral surface of heating elements;

θ –the temperature of heating elements;

θ_0 – the temperature inside the furnace;

dt – interval of elementary time;

c_u, c_a, c_z – mass heats (temperature-dependant) of the heated materials, attached elements, and crucible;

m_u, m_a, m_z – pieces weight, attached elements, and crucible;

$d\theta$ – elementary interval of temperature.

2.5.2 The achievement for a design algorithm for a solar furnace used in metallic material heating

Solar panel positioning by implementing the mathematical model needs the astronomic considerations.

In order to determine the real position of the sun on the sky, the following angles are important; θ_z – Zenith angle and γ_s Azimuth angle, as shown in **Figure 11**.

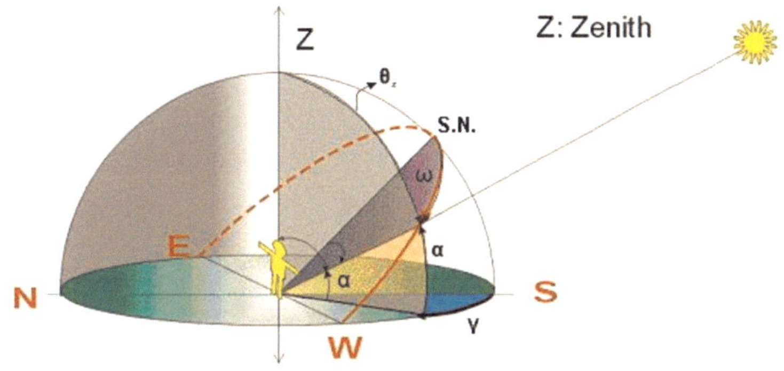

Figure 11.
Sun's trajectory on the sky – Important angles.

The calculus of these angles is made with mathematical formulas. Calculus formula for Zenith angle can be calculated by the relation:

$$\cos\theta_z = \sin\varphi \cdot \sin\delta + \cos\varphi \cdot \cos\delta \cdot \cos\omega$$

where **ϕ** is the latitude and is constant for the place where the solar tracker is positioned, for example, for Brasov, it is 45°39′, and δ is a declination and ω is an hour angle.

2.5.3 Virtual design of a heating solar furnace

In order to determine the minimum dimensions of the solar panel, a concave mirror is used as shown in **Figure 12**. The calculi are made on shading intervals as well

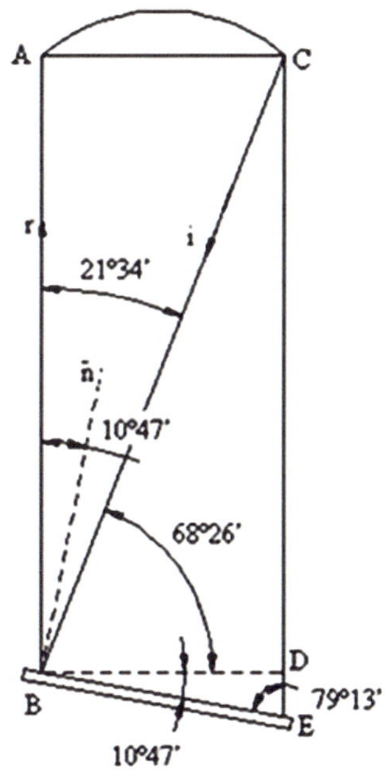

Figure 12.
The minimum dimensions of the solar panel.

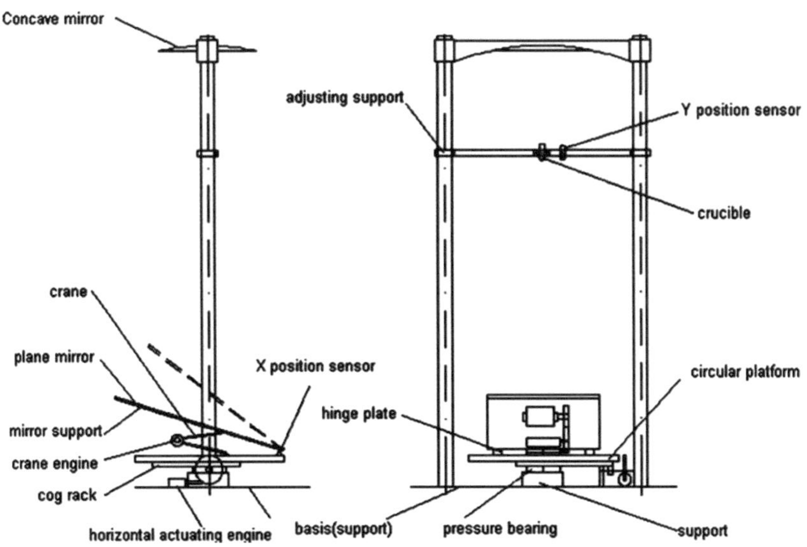

Figure 13.
The mirror position to create maximum thermal flow toward the furnace's crucible.

as on illumination optimum of the mirror in order to create maximum thermal flow toward the furnace's crucible (**Figure 13**).

Determination of minimum dimensions of the solar panel.

AC = Φ concave mirror.

$\hat{C} = 90–21.34 = 68.26.$

$AC = \sin \hat{B} \bullet BC \rightarrow BC = \frac{AC}{\sin \hat{B}} \rightarrow BC = \frac{AC}{0.363}$

$AB = \sin \hat{C} \cdot BC \rightarrow AB = 0.928 \cdot \frac{AC}{0.363} \rightarrow AB = 2.558 \cdot AC$ (AB minim)

$BD = AC$

$BD = BE \cdot \sin \hat{E}$

$BE = \frac{BD}{\sin \hat{E}} = \frac{AC}{0.98} = AC \cdot 1.0183$

$BE_{\max(45^\circ)} = AC \cdot \sqrt{2} \cdot 1.15; (1.15 = \text{const.})$

$BE_{\min} = AC \cdot 1.0183$

2.6 The characterization of the melting/burning/purifying process in solar furnaces

These objectives consisted in the development of some mathematical formalism that can allow better exploitation of the advantages accomplished during melting/burning/purifying processes in solar furnaces. The documentary stages made up until now are useful in the elaboration of these formalisms.

2.6.1 The achievement of a design algorithm for a solar furnace used in melting/burning/purifying of the metallic and nonmetallic materials

The solar furnace is a laboratory experimental furnace. During the experiments, we propose to process small quantities of material until 1 kg.

a. **Crucible dimensioning**

In order to make the crucible, graphite material is chosen due to its high temperatures characteristic until 2000°C and the relative high thermal conductivity. For accomplishing the calculation, aluminum is chosen as the test material due to its density up to 1 kg more volume than other materials that will be processed.

The volume of the crucible is calculated using the relation:

$$v = \frac{m}{\rho'} = \frac{\pi \cdot d^2}{4} \cdot h \quad (9)$$

where ρ' is the material density at the ambient temperature of 20°C, for example, aluminum has $\rho'_{met\_topit} = 2.72\ \mathrm{kg/dm^3}$.

b. **Dimensioning of the furnace masonry and thermal calculus**

The furnace has a cylindrical shape. Its disposal is vertical, and on its walls it chose a configuration with multiple layers, considering the temperature, as shown in **Figure 14**:

Let it be the exterior temperature of the refractory coat of 1200°C.

c. **Resistors**

The choice of the material from which the resistors are made should be considered so that the maximum working temperature exceeds almost 2–10% of the maximum temperature of the furnace. Knowing that the temperature within the furnace can reach 1600°C, Kanthal is chosen for resistors. Kanthal resistors are like spiral wires.

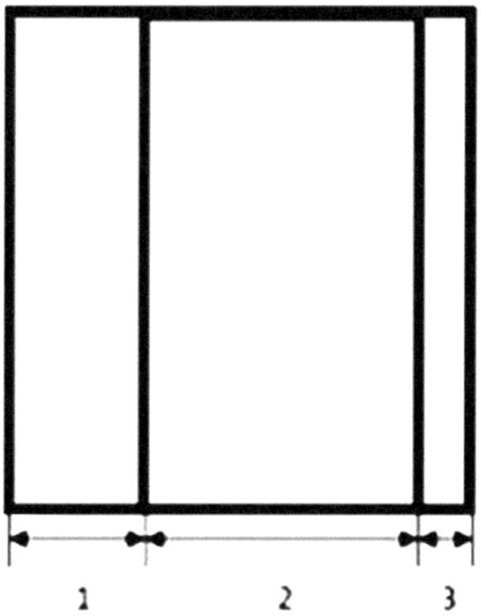

Figure 14.
The diagram of furnace wall: 1 – Refractory coat made of chromium magnesite with a thickness of 15 mm; 2 – Thermal insulation made of 700 diatomites treated with binders with a thickness of 30 mm; 3 – Exterior shell made of steel plate with a thickness of 2 mm.

In each chamber of the furnaces, 27 resistors are placed in the channels of the refractory material. They are disposed symmetrically on vertical and placed over the crucible so that we have a uniform temperature in the entire chamber of the solar furnace.

The length of each Kanthal spiral is 105 mm. The equivalent resistance on each chamber is 51.5 Ω. **Figure 2** shows the disposal of all 27 resistors.

The installed power of the furnace is calculated by the relation:

$$P_i = k \cdot P = k \cdot \frac{Q_i}{t_i} \qquad [W] \tag{10}$$

where k = 1.1÷1.5 is the safety coefficient that takes into account the possibility of forcing heating regime of the cold furnace, the possibility of decreasing network tension toward its nominal value, the possibility of decreasing time of the thermal insulation properties, the possibility of heating elements aging – that determine a high strength than the one initially calculated and in addition a smaller developed power (**Figure 15**).

The furnace with small dimensions, two chambers with a crucible, and the volume of 0.5 liters for each active chamber will be the first lab furnace. The reason the furnace is constructed with two working chambers is for the optimization of the working time.

We choose for a chamber furnace with crucible, because of its simple construction, the possibility of using it for different processes (for example, melting, burning, and purifying) as the possibility of realizing some different thermal regimes into the furnace, in essence, is what we propose to accomplish.

The resistor furnace (in the future, we will extend our research on an induction furnace too) has an alternate functioning regime because of the following functioning cycle:

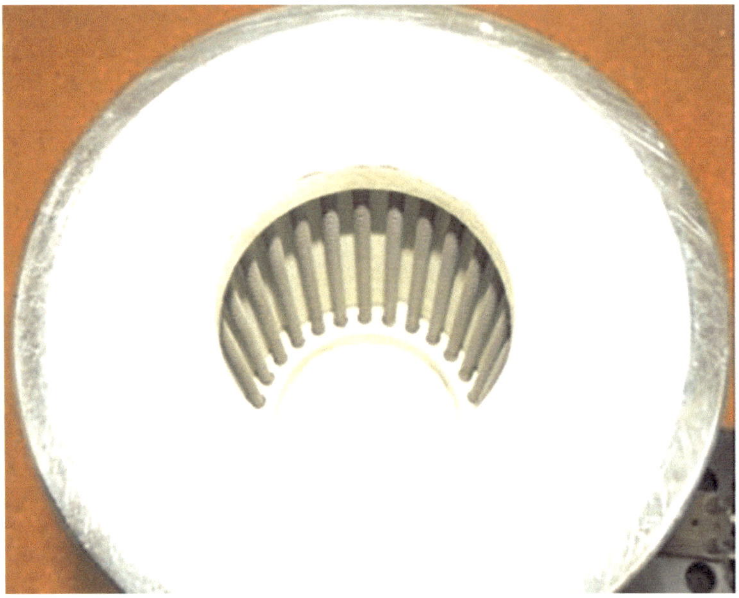

Figure 15.
Chamber disposal of the resistors.

- Loading of the crucible with the metal that will be processed;
- Heating;
- Melting;
- Burning;
- Purifying;
- Unloading.

The furnace will have the possibility of fitting in all working temperatures that are considered to be classifying criteria in heating electric heating technology:

- Low-temperature furnace (between 600 and 700°C the maximum value of the temperature);
- High-temperature furnaces (until 1600°C).

2.6.2 Virtual design of a melting/burning/purifying solar furnace

The scheme of the resistor furnace is presented in **Figure 16**. The components of the chamber furnace with crucible are as follows:

- chambers made of refractory material (1) and thermal insulation (2);
- the heating elements (3) are placed on the lateral walls of the furnace;
- crucible (4);
- material that will be processed (5);

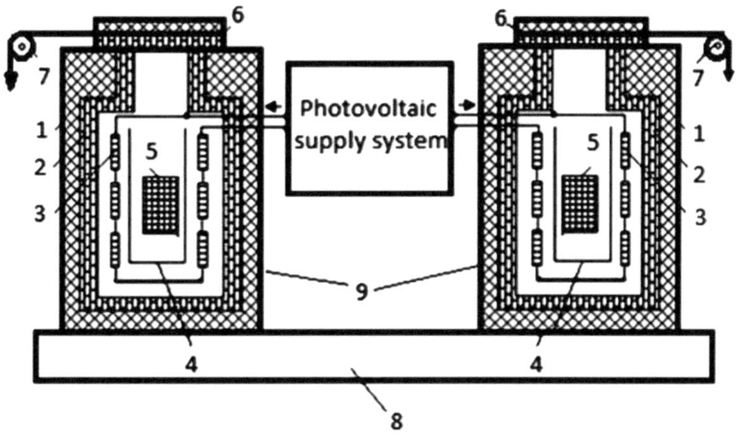

Figure 16.
Scheme of the furnace supplied from a photovoltaic system with resistors for materials melting/burning/purifying.

- the furnace presents the door (6), acted by the lifting device (7) where processed materials are introduced;
- furnace support (8);
- Metallic shell (9).

For the construction of the electrothermal furnaces, a series of specific materials are used, which are necessary for making furnace chamber, for the heating elements as well as for the measurement systems of the temperature.

3. Results and discussions

3.1 Identification of the optimum geometric arrangement of the components of a solar furnace for heating metallic materials

The photovoltaic system will be composed of (see **Figure 17**):

- solar panels (it chose panels made of polycrystalline silicon with a very good price/quality ratio and with a guarantee of 20 years) with the required energy;
- batteries system necessary for energy storage and furnace usage under adverse conditions in terms of solar radiation (in the evening or on cloudy days);
    - tension regulator;

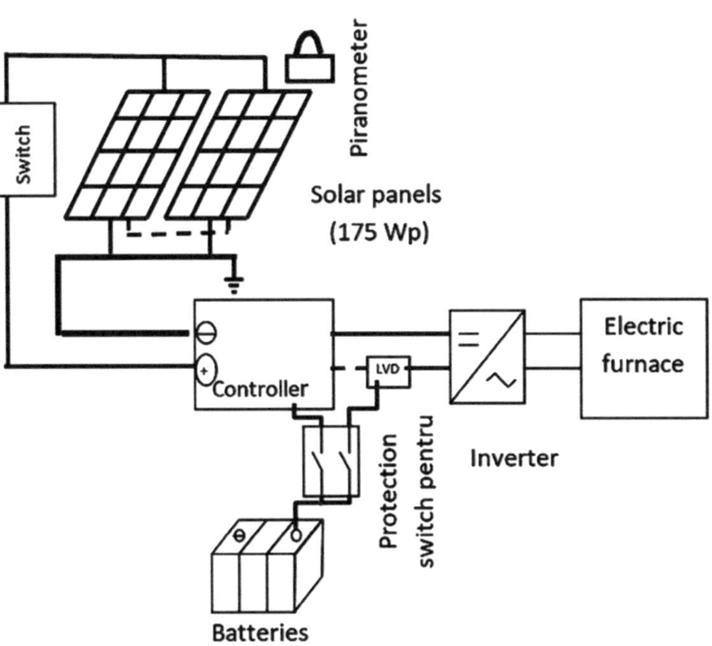

Figure 17.
The photovoltaic system necessary for supplying the solar furnace with resistors.

Figure 18.
Assembly images of the solar furnace.

- inverter necessary for transforming continuous current into the alternative current;

- accessories;

- solar tracker system (optional – necessary for obtaining a better efficiency of the photovoltaic system, which is permanently oriented so that the solar radiation would drop perpendicular on solar panels).

The temperature control in resistor furnaces has a special influence on the quality of the final products and on the specific energetic consumptions. With respect to the specific conditions of the technological process, especially the allowed temperature variations in the furnace and in the material, adjustment systems are used with intermittent action or a continuous one. The assembly image of the built solar furnace is given in **Figure 18**.

A PID control algorithm will be used for temperature control using PtRh-18 class thermocouple as a sensor. The maximum working temperature of these thermocouples is 1820°C [1, 2].

The PID algorithm is implemented using LabVIEW graphic programming language. **Figure 19** presents the panel, program interface, and PID application, and **Figure 20** presents the diagram and the proper program.

3.2 Interpretation of the achieved results

The solar furnace works by using electric energy produced by a photovoltaic system, which converts solar energy, solar radiation, into electric energy.

For a feasibility study for the solar radiation in the Brasov area, an SPN1 pyranometer is purchased. Global and diffuse solar radiation is measured using this device, and direct solar radiation can also be calculated using this device. Solar radiation monitoring starts from February. Global and diffuse solar radiation monitoring is noticed at every 5 minutes.

The conclusions that can be drawn after monitoring the solar radiation, according to **Figures 21–23**, are as follows:

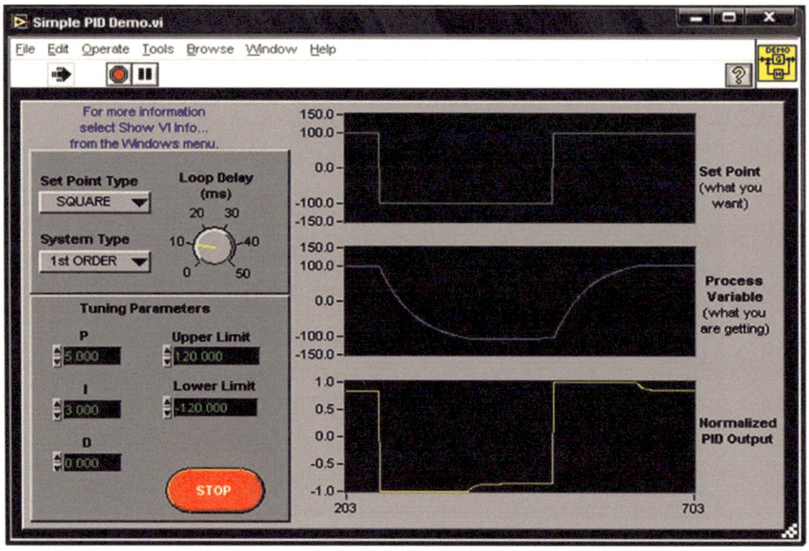

Figure 19.
The interface of the PID application.

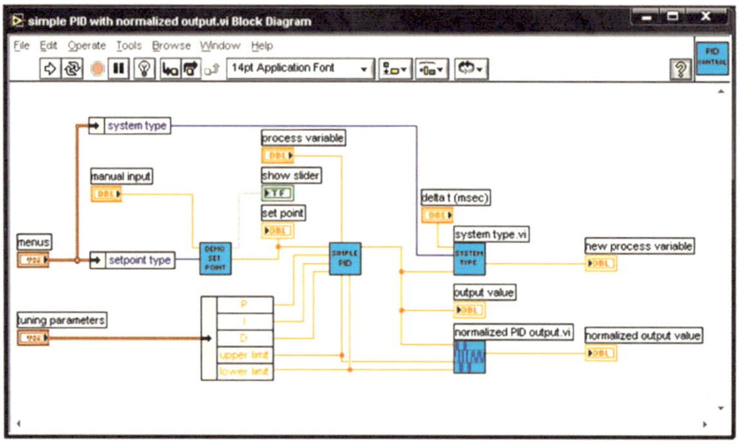

Figure 20.
Diagram of PID application.

- the maximum number of cloudy consecutive days was five, but in those days there were time intervals where horizontal global solar radiation exceeded the value of 400 W/m^2, the value that gives the possibility to the photovoltaic panel's system to offer enough energy so that the solar furnace with resistor would function under optimum conditions;

- the photovoltaic system is capable of producing almost 6.8 kW in a clear sky day (see **Figure 21**), thus allowing the furnace usage at maximum capacity and the possibility to store excess energy with the help of the solar batteries;

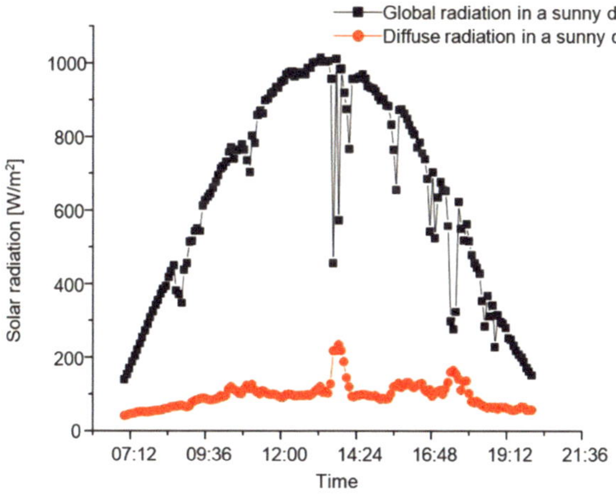

Figure 21.
Distribution of global and diffuse solar radiation during a sunny day (14 June).

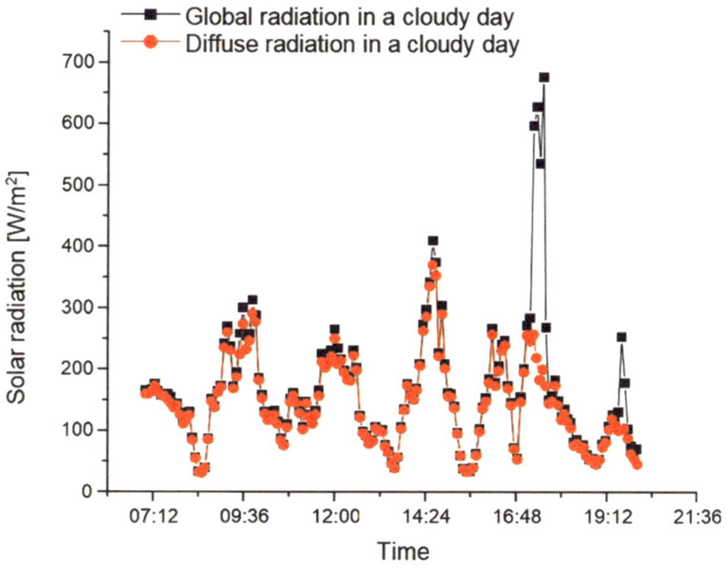

Figure 22.
Distribution of global and diffuse solar radiation during a cloudy day (12 June).

- During cloudy days, as shown in **Figure 22**, the quantity of energy generated by the system reduces considerably, so it only produces 1.5 kW, which would be enough for the judicious usage of the furnace;

- Yet, if there are long periods, as shown in **Figure 23**, then the system will need the energy to be stocked in the solar batteries; batteries system is designed to assure an autonomy of up to 5 days without raising the costs of the system very much;

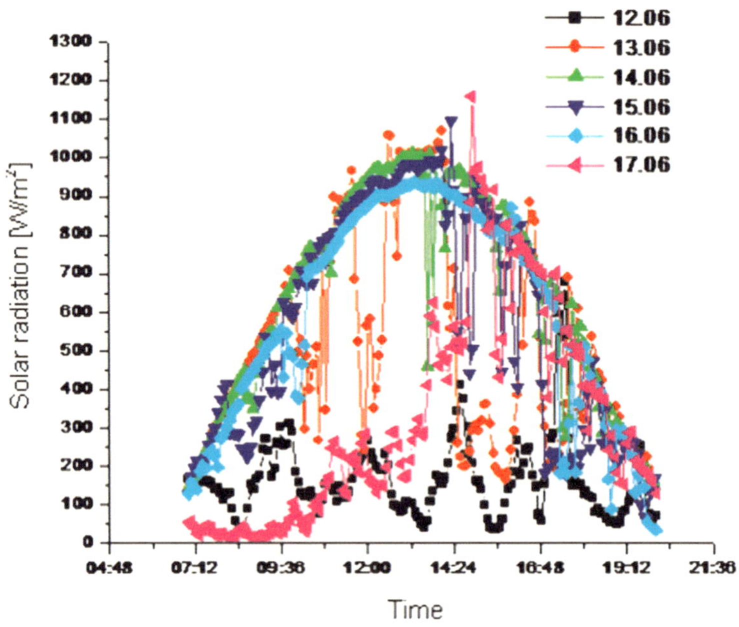

Figure 23.
Distribution of horizontal global radiation during the interval 12–17 June.

- For example, for the weakest days energetically speaking, 12 June, the energy quantity a system can generate is under 800 W.

4. Conclusions

The solar furnace can be used for numerous methods of technological purposes. The benefits that come from this sort of electrothermal installation is primarily based totally on sun electricity sources, and it could be utilized in various unique applications such as:

- For manufacturing metallic alloys;
- For heating operations of thermal processing (hardening and annealing, aging, carburizing, nitro-carburizing) and hot forming processing;
- For manufacturing abrasive materials, calcium carbide, and electrographite;
- For welding the metals and plastic packages;
- For manufacturing glass and ceramics materials;
- For nonmetallic products drying, preparation, or sterilization.

The researchers approach a dynamic thematic of high interest for problems consistency and for the wide area of applications.

The results already achieved within this stage are also noted with the help of the collaboration with researchers from the renowned research institute from Almerian Solar Platform, Spain.

Objectives' fulfillment required an ample information activity. It was necessary to have a good control of some different mathematical methods as well as good information regarding the experimental research in materials' melting/burning/purifying process.

The importance and complexity of this thematic make necessary detailed, interdisciplinary research. Thus, the team consists of specialists from the top domain of actual fundamental research such as materials science and physics.

The results already achieved open new work perspectives for the next year within the project in the approached thematic.

5. Perspectives

The materials that can be studied are quartz, zirconium dioxide, corundum, ceramic oxides and materials like carbides, nitrides, and boron for which conventional melting techniques present a series of inconvenient.

The practice possibility for using solar furnaces in steel melting can also be studied. The performances of such a solar furnace must not be special because there is sufficient temperature of 2000–2500°C.

In the future studies, there may be a higher interest for melting some other metals more expensive than steel like titanium, zirconium, and molybdenum. In this case, it must be assured an inert protective atmosphere, and thereby the complications and expenses related to these must be taken into account.

Other applications of solar furnaces are impurities evaporation, zonal melting, fractionates crystallization, zirconium oxide extraction from zircon (zirconium silicate), and the materials studied under thermal shock conditions.

Author details

Petrica Vizureanu*, Madalina-Simona Baltatu, Andrei-Victor Sandu, Dragos-Cristian Achitei, Dumitru-Doru Burduhos-Nergis and Manuela-Cristina Perju
Faculty of Materials Science and Engineering, "Gheorghe Asachi" Technical University, Iasi, Romania

*Address all correspondence to: peviz2002@yahoo.com

References

[1] Vizureanu P, Samoila C. The analysis of heating process in solar furnaces. Metalurgia International. 2008;**13**(2):5-10

[2] Vizureanu P. The analysis of the melting process of the materials in the solar furnaces. Metalurgia International. 2009;**14**(5):5-9

[3] Steinfeld A, Palumbo R. Solar Thermochemical Process Technology, Ency Phys Sci Techno. 3rd ed. MA: Academic Press Cambridge; 2003. pp. 237-256

[4] Roldan MI, Monterreal R. Heat flux and temperature prediction on a volumetric receiver installed in a solar furnace. Applied Energy. 2014;**120**:65-74

[5] Trefilov VI, Schur DV, Nagornaya NR. The solar furnaces for scientific and technological investigation. Renewable Energy. 1999;**16**(1–4):757-760

[6] Bjorndalen N, Mustafiz S, Islam MR. High temperature solar furnace: Current applications and future potential. Energy Sources. 2003;**25**(2):153-159

[7] Rodriguez J, Canadas I, Zarza E. New PSA High Concentration Solar Furnace SF40, Solar Paces, 2016, Solarpaces 2015: International Conference on Concentrating Solar Power and Chemical Energy Systems 1734

[8] Kovacik J, Emmer S, Canadas I. Solar furnace: thermal shock behavior of TiB2 coating on steel. In: 23rd International Conference on Metallurgy and Materials, Metal 2014: 23RD International Conference on Metallurgy and Materials. 2014. pp. 863-868

[9] Fernandez-Gonzalez D, Prazuch J, Verdeja LF. Solar synthesis of calcium aluminates. Solar energy. 2018;**171**:658-666

[10] Oliveira FAC, Rosa LG, Shohoji N. Nitriding VI-group metals (Cr, Mo and W) in stream of NH3 gas under concentrated solar irradiation in a solar furnace at PSA (Plataforma solar de Almeria). Solar Energy. 2015; **114**:51-60

[11] Ruiz-Bustinza I, Canadas I, Vazquez AJ. Magnetite Production from Steel Wastes with Concentrated Solar Energy. Steel Research International. 2013;**84**(3):207-217

[12] Flamant G, Ferriere A, Monty C. Solar processing of materials: Opportunities and new frontiers. Solar Energy. 1999;**66**(2):117-132

[13] Levenfeld B, Varez A, Rodriguez J. Study of the densification, mechanical and magnetic properties of Ni-Zn ferrites sintered in a solar furnace. J. Gutierrez-Lopez, Ceramics International. 2016;**41-5**:6534-6654

[14] Milosan I, Cristea D, Oancea G. Characterisation of EN 1.4136 stainless steel heat-treated in solar furnace. International Journal of Advanced Manufacturing Technology. 2019;**101** (9–12):2955-2964

[15] Ceballos-Mendivil LG, Carvajal-Campos Y, Estrada CA. Solar synthesis of nanostructured zirconia: microstructural and thermal characterization. Materials Research Express. 2020;7(11):115014

[16] Milosan I, Florescu M, Bedo T. Evaluation of Heat-Treated AISI 316 Stainless Steel in Solar Furnaces to Be Used as Possible Implant Material. Materials. 2020;**13**(2):581

[17] Catana D, Rodriguez J, Milosan I. Aspects of thermal transfer in heat

treatment of alloy steels using concentrated solar energy. Journal of Thermal Analysis and Calorimetry. 2019; **138**(4):2541-2553

[18] www.energosun.ro

Chapter 5

The Technical Challenges of the Gasification Technologies Currently in Use and Ways of Optimizing Them: A Review

Ali Mohammadi and Anthony Anukam

Abstract

Since the world is gradually drifting toward sustainable development, renewable energy technologies are gaining traction and gasification technology is one of many renewable energy technologies that have gained popularity in recent times. The gasification technology is one of three main (combustion and pyrolysis) thermochemical conversion pathways that can be used to recover energy from biomass materials. Although the gasification technology has been in existence for centuries, it has not been exploited to its full potential mainly because the fundamental principles underpinning its operation are still vague, particularly with regard to feedstock flexibility and the type of gasification system. Furthermore, due to the many types of gasification systems, the mechanisms involved in their feedstock conversion processes are still under debate and require further research to clearly establish the optimum conditions of performance of each type of gasifier. Therefore, this chapter presents an overview of the gasification technology and discusses the different types of gasification systems that are commonly used today for the recovery of energy. The limitations of each type of gasifier in relation to performance and feedstock conversion are also discussed, including research priority areas that will allow for system optimization in terms of efficiency.

Keywords: energy recovery, gasifiers, combustion kinetics, feedstock, gasification efficiency

1. Introduction

Evidence suggests that conventional energy production has limited capacity to meet growing demand and that additional demands will have to be met by unorthodox sources. Since the world is now drifting toward sustainable development, renewable energy technologies are gaining traction. One of such renewable energy technologies that has received great attention in recent times include biomass gasification, which is one of three main (combustion and pyrolysis) thermochemical conversion pathways used to recover energy from biomass materials. Gasification produces energy from biomass and involves heating the biomass at elevated temperatures (above 1000°C) under a limited supply of oxygen to produce a mixture of gases (H_2, CO, CO_2) collectively referred to as syngas. However, the combustible constituents of the syngas are CO and H_2 and can be used as fuel in gas engines for heat and electricity generation as well as for the production of chemicals (such as alcohols, organic acids, ammonia, and methanol) via the Fischer-Tropsch process [1]. The significance of gasification technology is such that it helps waste management, at the same time, produces energy and other valuable products needed for economic growth. Systems designed to gasify coal is assumed to be able to use biomass as well, however, differences in the characteristics of coal and biomass can have a significant impact on the sizing and design of the combustion chamber of the gasification system, as well as on the location of the gasifying agent [2]. A graphical representation of a gasification process, which depicts feedstock flexibility and the production of a wide range of products, is presented in **Figure 1**.

The gasification technology has existed for several decades and has, as of today, been commercialized in very few countries of the world like Sweden, Germany, Canada, the United States, India, and China. The use of this technology offers a number of ecological and economic advantages such as low emission of pollutants, reduction in the environmental effects of waste disposal, generation of non-hazardous by-products when biomass is used as the feedstock, and lower operating cost [4].

Gasification occurs in a gasifier under a series of chemical reactions that are mostly endothermic in nature; however, to provide the heat required for the reactions to proceed successfully, and the heat needed for drying and pyrolysis to occur, a certain amount of exothermic combustion is allowed in the gasifier [4, 5]. The gasification reactions are described in greater detail in subsequent sections. The gasifier and its configuration are key factors that affect the entire gasification process, including the reactions occurring and their products [6]. This is true because gasifiers are generally classified into three broad groups, namely: the fixed bed gasifiers, the fluidized bed

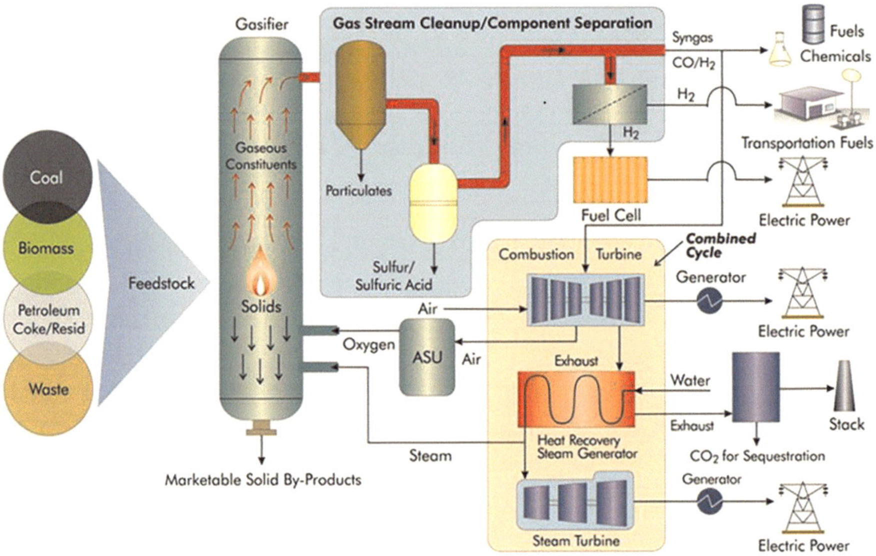

Figure 1.
A schematic representation of a gasification process depicting feedstock flexibility and the wide range of products that can be obtained from the process [3].

Type of gasifier	Characteristics
The fixed beds	1. Small capacity gasifiers (typically from 0.01–10 MW) 2. Can handle large and coarse particles 3. Low product gas temperature (450–650°C) 4. High particulate content in the gas product stream 5. High gasification agent consumption 6. Ash is removed as slag or dry 7. May result in high tar content (0.01–150 g/Nm^3)
The fluidized beds	1. Medium capacity (1–100 MW) 2. Uniform temperature distribution 3. Better gas-solid contact 4. High operating temperature (1000–1200°C) 5. Low particulate content in the gas stream 6. Suitable for feedstocks with low ash fusion temperature 7. Ash is removed as slag or dry
The entrained flows	1. Large capacity (60–1000 MW) 2. Needs finely divided feed material (0.1–0.4 mm) 3. Very high operating temperatures (1200°C) 4. Not suitable for biomass feedstocks 5. Very high oxygen demand 6. Short residence time 7. Ash is removed as slag 8. May result in low tar content (negligible)

Table 1.
The main characteristics of the three types of gasifiers commonly used for the recovery of heat and electricity from biomass [7].

gasifiers, and the entrained flow gasifiers. **Table 1** shows the main characteristics of these three gasifiers.

Although the gasification technology may be considered as a useful technology for the recovery of energy from biomass materials, the technological choices with regards to the type of gasification system (fixed bed, fluidized beds, or entrained flow reactors) for the conversion of biomass are still faced with a host of technical barriers that have hindered the significant exploitation of the gasification technology and biomass energy as a whole. The quality of the syngas produced from the gasification process, the lack of feedstock flexibility and its mechanism of conversion are the main obstacles. This chapter, therefore, presents an overview of the gasification technology and discusses its main technical barriers with reference to the gasification systems commonly used today. The status of current research in gasification and future research focus are also presented.

2. Types of gasification systems and their configurations

There are different types of gasification systems but the most commonly used are the fixed-bed, fluidized-bed, and entrained-flow gasification systems. The main differences between these gasifiers are connected to their mechanism of heating and the way feedstock and gasifying agents are introduced in the gasification process, as well as by the location of syngas output [8–10]. However, the technological choices toward these gasifiers are guided by the nature and availability of biomass feedstocks. While

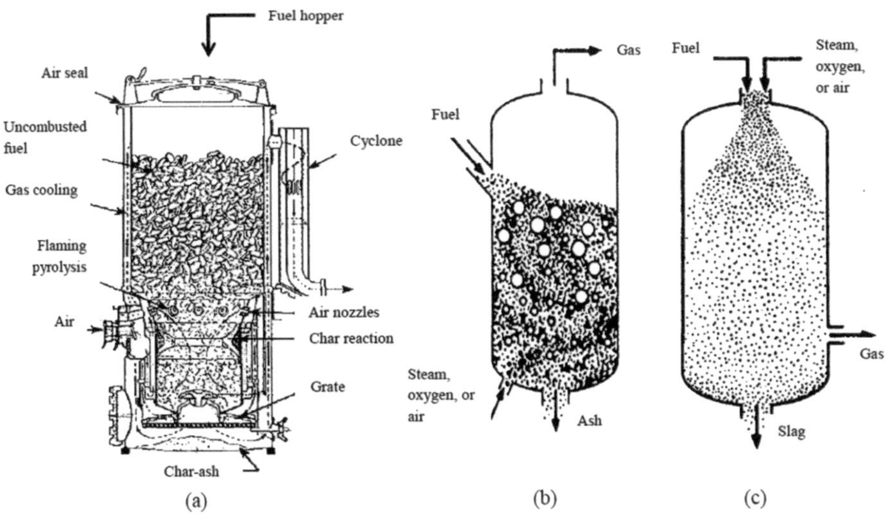

Figure 2.
Schematic representations of the gasification systems in use today: (a) fixed bed; (b) fluidized bed; (c) entrained flow. Reproduced with permission from [13, 15].

the characteristics of biomass feedstocks intended for gasification are detailed in [11], the principles of operation of the types of gasifiers mentioned above and their merits and demerits are equally well described in [12, 13] and in [14]. These gasification systems may appear as simple devices but their successful operations are not so simple. The gasifiers are still faced with a host of technical issues that have hindered their broader market penetration. These technological barriers are described in Section 5. Nonetheless, in order to fully comprehend the technical barriers of each of these gasifiers, it is important to understand the differences between the gasifiers in terms of configuration, which also affects the thermodynamics of their operation. Therefore, a schematic diagram of each gasifier type is presented in **Figure 2**.

3. The gasification reaction chemistry

The key mechanism of the gasification technology involves the conversion of solid carbonaceous materials like biomass into flammable gas by partial oxidation. However, the chemistry involved in the process is quite complex and can be achieved via a series of physical and chemical transformation reactions that occur inside the gasification system [4, 16]. The major chemical reactions occurring are those that involve the degradation of large organic molecules into carbon monoxide (CO), carbon dioxide (CO_2), hydrogen (H_2), water in the form of steam (H_2O), and methane (CH_4). These reactions take place in accordance with the chemical bonding theory and can be represented thus [3]:

The combustion reactions include:

$$C + \tfrac{1}{2}O_2 \rightarrow CO\ (-111\ \text{MJ/kmol}) \quad (1)$$

$$CO + \tfrac{1}{2}O_2 \rightarrow CO_2\ (-283\ \text{MJ/kmol}) \quad (2)$$

$$H_2 + \tfrac{1}{2}O_2 \rightarrow H_2O\ (-242\ \text{MJ/kmol}) \quad (3)$$

Other key gasification reactions are:

$$C + H_2O \leftrightarrow CO + H_2 \ (+131 \text{ MJ/kmol}) \tag{4}$$

$$C + CO_2 \leftrightarrow 2CO \ (+172 \text{ MJ/kmol}) \tag{5}$$

$$C + 2H_2 \leftrightarrow CH_4 \ (-75 \text{ MJ/kmol}) \tag{6}$$

The above reactions occur under standard operating conditions of gasification and are considered important reactions that form the major part of the syngas produced in the gasification process [4, 16]. While reaction (4) may be referred to as the "Water-Gas Reaction", reactions (5) and (6) are termed the "Boudouard Reaction" and the "Methanation Reaction" respectively. Reactions (4) and (5) are the main reduction reactions. However, under high carbon conversion conditions, reactions (4)–(6), being heterogeneous in nature, are reduced to the following homogeneous gas-phase reactions [16]:

$$CO + H_2O \leftrightarrow CO_2 + H_2 \ (-41 \text{ MJ/kmol}) \tag{7}$$

$$CH_4 + H_2O \leftrightarrow CO_2 + 3H_2 \ (+206 \text{ MJ/kmol}) \tag{8}$$

Reactions (7) and (8) are known respectively as the "Water-Gas-Shift Reaction" and the "Steam-Methane-Reforming Reaction". These two reactions (7) and (8) play a key role in determining the final equilibrium of the composition of the syngas produced in the gasification process [3, 16]. Under a limited supply of oxygen to the gasifier, the sulfur composition of the feedstock is converted to hydrogen sulfide (H_2S), with a minute amount forming carbonyl sulfide (COS). The nitrogen (N) chemically bound in the feedstock is converted to gaseous nitrogen (N_2), ammonia (NH_3), and traces of hydrogen cyanide (HCN). The chlorine in the feedstock is mainly converted to hydrogen chloride (HCl). It is important to however state that the concentrations of sulfur, nitrogen and chloride in the feedstock for gasification are sufficiently low that their effects in the gasification process are quite insignificant; trace elements (such as arsenic, mercury, and other heavy metals) that are associated with both the organic and inorganic components of the feedstock are mostly contained in the fractions of ash and slag formed during gasification, as well as in the gases emitted, and must be expunged from the syngas prior to further use [16].

3.1 The kinetics of gasification reactions

Temperature increases in a gasification process lead to dehydration, volatilization, and degradation of the biomass feedstock. The gasification process reactions are mostly reversible reactions. The order of the reactions and their conversion rates are often subject to the limitations of the reaction kinetics and thermodynamic equilibrium of the gasification process. For instance, reactions (1)–(3) presented in a previous section are combustion reactions that actually go to completion when equilibrium positions of the reactions shift to the right. However, not all reactants in a gasification process can be completely converted into products; as such, stoichiometric calculations may be required to determine the products of a completed reaction [5].

While the kinetics of a reaction can determine how fast products are formed and whether the reactions in the gasifier go to completion, the equilibrium state of the reaction determines to what extent the reaction can progress. The thermal efficiency of the gasification process and the composition of the syngas produced are strongly influenced by the thermodynamic equilibrium of the water-gas-shift reaction and the steam-methane-reforming reaction (reactions (7) and (8), Section 3) [3, 4]. A useful tool for evaluating important design parameters of a gasification technology is thermodynamic modeling. With this tool, process efficiency can be optimized at different operating conditions; the relative quantities of gasifying agents such as oxygen and steam can also be calculated including the composition of the product syngas.

4. Controlling factors to the stability of gasifier operation

An understanding of the technical challenges of gasification technology requires a basic understanding of the factors that control the stability of gasifier operation. A typical gasification process includes the following four key steps: drying, pyrolysis, oxidation, and reduction. There are no strict boundaries between these steps; they often overlap and a host of factors including the type of gasifier, feedstock type, and process parameters, such as temperature, determines the output of a gasification process involving the four key steps listed above [4, 5, 8]. The operating temperature of a gasification process is a function of the amount of oxygen fed to the gasification system (gasifier), which induces partial gasification. Temperature response will abruptly change at an equivalence ratio (ER) of about 0.25; depending on the source of oxygen, this change point is typical of gasifier temperatures in the range 600–800°C; some quantities of oil and tar are produced in the pyrolysis stage of the gasification process. These products of the pyrolysis stage are stable for about a second at temperatures lower than 600°C [13].

The fixed bed updraft gasifier operates at temperatures below 600°C and generates considerable amounts of tars that are often emitted with the syngas, while its counterpart, the downdraft gasifier (also of the fixed bed type) is self-regulating and produces far less tar relative to the updraft gasifier; the fluidized bed gasifier also has high tar production rate, in fact, its tar production rate is greater than other types of gasifiers like the fixed bed and the entrained flow gasifiers [13, 14, 17].

5. Technical challenges of the gasification technology

Although the gasification technology has experienced development over several decades and has been commercialized in a number of countries like those previously mentioned [14]; its successful operation is not as simple as can be imagined because of the thermodynamics of the operation of the technology are not well understood. Further exploitation of the technology still needs to overcome a considerable number of technical issues. A description of the technological barriers that are associated with each type of gasification technology is presented thus:

5.1 Limitations of the fixed bed gasifier

The fixed bed gasifiers (updraft, downdraft, and crossdraft) are the simplest of all the types of gasifiers and are mainly suitable for small-scale applications (<10 MWth) [5]. Although they (fixed bed gasifiers) are very advantageous in terms of their simplicity and ease of operation, they generally suffer from poor mixing and poor heat transfer within the gasifier, which makes it difficult to achieve even distribution of fuel and temperature across gasifier geometry hence scale-up of this type of gasifier is difficult. The fixed bed downdraft gasifier, which has not satisfactorily performed with feedstock capacity beyond 425 kg/h [18], is a typical example of the described technical issue. This is because air cannot travel up the center of the gasifier, which creates cold spots in and around the combustion zone of the gasifier during operation and results in reduced gasification efficiency. This limitation has been attributed to design characteristics in terms of gasifier geometry (throat angle and throat diameter) and air inlet velocity. In the case of the updraft gasifier however, its high tar production rate (5–20%) [19] remains a challenge to date and renders this type of gasifier unsuitable where a clean product gas (syngas) is desired. Due to its high tar production rate, the updraft gasifier is well-suited for the gasification of low-volatile feedstocks like charcoal [5].

5.2 Limitations of the fluidized bed gasifier

In terms of configuration, the fluidized bed gasifier (FBG) operates on the principle of fluidization where a gas stream is forced through a particle bed vessel that behaves like a fluid under certain conditions such as high particle flow velocity. The commonly used fluidization media include air, steam, or mixtures of steam/oxygen. The FBG is the most efficient of all types of gasifiers and its efficiency is mainly dependent upon the thermochemical and fluid behavior inside the gasifier; this type of gasifier is more appropriate for medium-scale units of about 5–100 MWth [5, 20, 21]. From a system performance and technical point of view, the operation of the FBG is quite complex because of the need to simultaneously control air supply, bed material, and feedstock during operation of the gasifier. As a result, the product gas obtained from the gasification process may be very high in particulates, which can circulate and cause equipment erosion. Although it may sway the gasification process, the FBG is operated at high-pressure conditions, which can result in low volumetric gas flow rates, condensation during compression, and other operational complications such as defluidization from particle agglomeration particularly when agricultural crops and wastes are used as feedstock in the gasification process. This is because agricultural crops and wastes contain an increased amount of ash/alkali and, the alkali content of ash (such as sodium and potassium alkali) can form low-melting eutectics with the silica in the sand, which is the regularly used bed material in FBG processes [22]. Under this condition, agglomeration and sintering will occur, triggering the formation of a thin sticky substance around the bed particles with an instant loss of bed fluidization (defluidization). Typical factors influencing agglomeration and the loss of fluidization in FBGs are presented in **Table 2**.

Even if more sophisticated bed materials such as alumina and magnesite are used in the FBG process of feedstocks with high ash/alkali content, process cost will become an issue of concern. These types of technical issues call to question the feedstock flexibility of the FBG systems.

5.3 Limitations of the entrained flow gasifier

The entrained flow gasifier (EFG) is an old alternative energy production technology used on a large-scale (>50 MWth) [5] in the petroleum industry for the gasification of petroleum residues. This type of gasifier offers greater rates of collision between solid particles and is considered excellent in terms of performance because of vigorous mixing of feedstock and oxidizing agent as well as better feed conversion efficiencies in comparison to other types of gasifiers [25]. However, even though the

Parameter	Agglomeration and loss of fluidization (defluidization)
Temperature	The possibilities of agglomeration and defluidization are exacerbated by rising temperatures.
Steam	Agglomeration and defluidization can occur upon increase in steam during gasification due to the formation of molten sodium disilicate, which can occur via liquid-solid reaction under steam application conditions.
Alkalis, iron sulfides, and siderite	Increases the possibilities of the formation of sticky substances, which can, in turn, facilitate agglomeration and defluidization.
Fluidization velocity	The tendencies of agglomeration are lowered below the sintering temperature of ash when the velocity of fluidization is increased. The force of segregation also increases under this circumstance.
Particle size distribution	The possibilities of agglomeration and defluidization are high when bimodal or multimodal particle size distribution occurs.

Table 2
The summary of the impact of operating parameters on agglomeration and loss of fluidization [23, 24].

EFG has been in existence for centuries, it has not been exploited to its full potential partly because the fundamental principles underpinning its operation are still vague, particularly with regards to the type of material suitable as feedstock. The mechanisms involved in the feedstock conversion process are still under debate. In addition, the EFG is operated at very high temperatures (1,200 – 2,000°C) and pressures, under these operating conditions, fuel-oxygen mixtures are turned into a turbulent flame of dust that ensures the production of liquid ash, which are deposited on gasifier walls. This constitutes a technical issue of concern, particularly when analyzing the ash melting behavior of the material used as feedstock in the gasification process. Due to this high operating pressure, numerical modeling and experimental validation of the EFG tend to be onerous. Furthermore, due to its operating conditions, only specific types of materials are used as feedstock.

6. Current research status

Extensive studies have been undertaken on gasification technology over the last decade. Despite the numerous studies, however, there are still pending research-related issues (such as those described in preceding sections) that require further improvements. For example, Kaushal et al. [26] developed a one-dimensional steady-state model specific to the bubbling fluidized bed gasifier (BFBG). Gómez-Barea et al. [27] also reviewed the performance optimization of a small-scale FBG plant with the aim of maximizing char conversion rate and minimizing secondary gas treatments. The process performance of the downdraft gasifier was evaluated by Biagini et al. [28] in which the performance parameters such as syngas production, syngas heating value, cold gas efficiency, and the net efficiency of the gasifier were monitored using corn cobs as feedstock. Furthermore, the performance of a pilot-scale pressurized entrained-flow (EFG) plant using stem wood made from pine and spruce as feedstocks was assessed by Weiland et al. [29]. A combined system involving gasification, hydrothermal carbonization (HTC), and solid oxide fuel cell (SOFC) technologies was developed by Papa et al. [30] using commercial process simulation software (ASPEN Plus), where the focus was to investigate the efficiency of the system under various operating conditions. The challenges and opportunities of modeling the gasification technology using Aspen Plus were also detailed by Mutlu and Zeng who alluded to the issues of the gasification technology as hindering the widespread commercialization of the technology [31].

7. Research priority areas and solutions to the identified technical issues

The FBGs such as the downdraft gasifier is characterized by four distinct reaction zones including the drying, pyrolysis, combustion, and reduction zones respectively; the specific functions of each of these zones are described in [32]. Of these distinct reaction zones, the combustion zone, also known as the oxidation zone, is considered the most important zone because heat is generated in this zone. However, the presence of cold spots (a factor linked to uneven heat distribution in and around the combustion zone of the downdraft gasifier), is the main reason why these types of gasifiers are limited to small-scale applications [13]. There are basically two methods that can provide a solution to the problem of uneven heat distribution in fixed bed systems: one method is to decrease the cross-sectional area of the gasifier at a certain height. This means altering the design characteristics of the throat angle and throat diameter of the gasifier by way of size-reduction. The other method is to centralize the air inlet and its velocity using nozzles that are positioned in a way that allows the throat circumference of the gasifier to be captured.

In the case of the FBGs, although a well-established technology (in terms of design concept) for heat and power generation, bed defluidization, as indicated in a previous section, is considered the main technical issue, which as previously described, occurs due to agglomeration and pressure drops, particularly when gasifying feedstocks with high amounts of ash such as agricultural residues and wastes. Alkali silicates such as calcium, potassium, and sodium silicates present in ash can form low-melting eutectics with silica, which is often used as the bed material in FBGs [22]. A quick and easy solution to the defluidization problems in FBGs is to replace the commonly used bed material (silica) with more advanced artificial materials such as aluminum oxide or magnesium carbonate. However, the cost associated with the use of these materials may constitute a major drawback. Therefore, the hydrodynamics of the FBG needs to be further investigated and the hydrodynamic study must incorporate devolatilization kinetics, char gasification, and gas species in relation to particle agglomeration and sintering.

For the high-pressure EFG, the production of molten ash (which mostly originates from the ash constituents of the feedstock and forms deposits on the walls of the gasifier) is a commonly encountered technical problem. Depending on the operating conditions of the gasification process, the molten ash deposits often solidify, causing plugging and the blockage of critical parts of the gasifier thereby hindering process efficiency. Therefore, just like the FBG, a solution to the problem of molten ash formation in the EFG is to further investigate the feedstock conversion mechanism and gasifier hydrodynamics, particularly when more complex low-grade feedstocks such as agricultural residues and biomass-based chars are used in the gasification process under high-pressure conditions.

Studies [33, 34] have shown that modeling work has accelerated the research progress made in the field of biomass gasification since gasifier design and operating conditions can be optimized at minimal time and costs. However, modeling and simulation cannot replace good experimental investigations. In fact, studies [35] have determined that mathematical modeling and simulation of high temperature and pressure reaction systems involving gaseous, liquid, and solid phases is a major scientific challenge. Therefore, addressing the technical issues of the gasification technologies described in this chapter will not only require the development of a robust and sophisticated model that can be applied to a wider range of operating parameters of the gasifiers but also able to replicate actual operations of the gasification technologies with an acceptable level of anomaly.

8. Conclusions

Gasifier design and process optimization for complex biomass feedstocks, in general, are very challenging due to the lack of detailed understanding of the various thermochemical reaction steps governing the conversion of biomass feedstocks under high temperature and pressure conditions. The gasification technologies described in this chapter are multiphase systems that are characterized by complex operational steps. Therefore, in order to better comprehend the complex interactions between process steps during gasification and to address the technological issues earlier described, experimental studies under systematic variation of feed specification and process parameters are required. It is also necessary to ensure proper process mapping based on experimental data from lab- to pilot-scale in order to develop a comprehensive gasification process understanding and to provide a thorough data basis for the validation of numerical simulations. This implies the development of state-of-the-art experimental techniques that are applicable under the acrid conditions of gasification technologies.

Conflict of interest

The authors declare no conflict of interest.

Author details

Ali Mohammadi[1*] and Anthony Anukam[2]

1 Department of Engineering and Chemical Sciences, Karlstad University, Karlstad, Sweden

2 Biochemical Process Engineering, Division of Chemical Engineering, Department of Civil, Environmental and Natural Resources Engineering, Luleå University of Technology, Sweden

*Address all correspondence to: ali.mohammadi@kau.se

References

[1] Song X, Guo Z. Technologies for direct production of flexible H_2/CO synthesis gas. Energy Conversion and Management. 2006;**47**:560-569

[2] Ciolkosz D, Miller B, Wallace R. Renewable and Alternative Energy Fact Sheet: Characteristics of Biomass as a Heating Fuel. Pennsylvania, United State of America: The Pennsylvania State University, Ag Communications and Marketing; 2010

[3] National Energy Technology Laboratory. Gasification introduction. 2021. Available from: https://netl.doe.gov/research/Coal/energy-systems/gasification/gasifipedia/intro-to-gasification [Accessed: October 27, 2021]

[4] Anukam A, Mamphweli S, Meyer E, Okoh O. Computer simulation of the mass and energy balance during gasification of sugarcane bagasse. Journal of Energy. 2014:1-9

[5] Basu P. Biomass Gasification, Pyrolysis and Torrefaction: Practical Design and Theory. London, United Kingdom: Academic Press; 2018

[6] Arena U. Process and technological aspects of municipal solid waste gasification. A review. Waste Management. 2012;**32**:625-639

[7] Zhang Y, Cui Y, Chen P, Liu S, Zhou N, Ding K, et al. Gasification technologies and their energy potentials. In: Sustainable Resource Recovery and Zero Waste Approaches. Amsterdam, The Netherlands: Elsevier; 2019. pp. 193-206

[8] Pohjakallio M, Vuorinen T, Oasmaa A. Chemical routes for recycling—dissolving, catalytic, and thermochemical technologies. In: Plastic Waste and Recycling. Amsterdam, The Netherlands: Elsevier; 2020. pp. 359-384

[9] Anukam A, Mamphweli S, Reddy P, Okoh O, Meyer E. Assessment of the influence of torrefied biomass physical characteristics, design and operating variables on gasification efficiency. Journal of Chemistry. 2016;**10**:283-292

[10] Pickett MM. Modeling the Performance and Emissions of British Gas/Lurgi-Based Integrated Gasification Combined Cycle Systems. North Carolina, United State of America: North Carolina State University Libraries; 2001

[11] Anukam A, Berghel J. Biomass pretreatment and characterization: A review. In: Biomass. London, United Kingdom: IntechOpen Limited; 2020

[12] Anukam A, Mamphweli S, Reddy P, Meyer E, Okoh O. Pre-processing of sugarcane bagasse for gasification in a downdraft biomass gasifier system: A comprehensive review. Renewable and Sustainable Energy Reviews. 2016;**66**:775-801

[13] Reed TB, Das A. Handbook of Biomass Downdraft Gasifier Engine Systems. Colorado, United States of America: Solar Energy Research Institute; 1988

[14] Luo X, Wu T, Shi K, Song M, Rao Y. Biomass gasification: An overview of technological barriers and socio-environmental impact. In: Gasification for Low-Grade Feedstock. London, United Kingdom: IntechOpen Limited; 2018

[15] Breault RW. Gasification processes old and new: A basic review of the major technologies. Energies. 2010;**3**:216-240

[16] Higman C, van Der Burgt M, Higman C, Vanderburgt M. The thermodynamics of gasification. In: Gasification. Amsterdam, Boston: Gulf Professional Pub./Elsevier Science; 2008. pp. 11-31

[17] Bhatia S. Advanced Renewable Energy Systems (Part 1 and 2). London, United Kingdom: CRC Press; 2014

[18] Mamphweli NS. Implementation of a 150KVA Biomass Gasifier System for Community Economic Empowerment in South Africa. 2009

[19] Desrosiers R. Fundamental air-gasification engineering parameters. In: Annual Report for Fiscal Year 1981. Golden, CO (USA): Solar Energy Research Inst; 1982

[20] Olofsson I, Nordin A, Söderlind U. Initial Review and Evaluation of Process Technologies and Systems Suitable for Cost-Efficient Medium-Scale Gasification for Biomass to Liquid Fuels. Umeå, Sweden: Umeå Universitet; 2005

[21] Pfeifer C, Koppatz S, Hofbauer H. Steam gasification of various feedstocks at a dual fluidised bed gasifier: Impacts of operation conditions and bed materials. Biomass Conversion and Biorefinery. 2011;**1**:39-53

[22] Heyne S, Liliedahl T, Marklund M. Biomass Gasification-A Synthesis of Technical Barriers and Current Research Issues for Deployment at Large Scale. 2013

[23] Kuo J-H, Lin C-L, Wey M-Y. Effect of alkali concentrations and operating conditions on agglomeration/ defluidization behavior during fluidized bed air gasification. Powder Technology. 2011;**214**:443-446

[24] Lin C-L, Peng T-H, Wang W-J. Effect of particle size distribution on agglomeration/defluidization during fluidized bed combustion. Powder Technology. 2011;**207**:290-295

[25] Unar IN, Wang L, Pathan AG, Mahar RB, Li R, Uqaili MA. Numerical simulations for the coal/oxidant distribution effects between two-stages for multi opposite burners (MOB) gasifier. Energy Conversion and Management. 2014;**86**:670-682

[26] Kaushal P, Abedi J, Mahinpey N. A comprehensive mathematical model for biomass gasification in a bubbling fluidized bed reactor. Fuel. 2010;**89**: 3650-3661

[27] Gómez-Barea A, Leckner B, Perales AV, Nilsson S, Cano DF. Improving the performance of fluidized bed biomass/waste gasifiers for distributed electricity: A new three-stage gasification system. Applied Thermal Engineering. 2013;**50**: 1453-1462

[28] Biagini E, Barontini F, Tognotti L. Gasification of agricultural residues in a demonstrative plant: Corn cobs. Bioresource Technology. 2014;**173**: 110-116

[29] Weiland F, Hedman H, Marklund M, Wiinikka H, Öhrman O, Gebart R. Pressurized oxygen blown entrained-flow gasification of wood powder. Energy & Fuels. 2013;**27**:932-941

[30] Papa AA, Di Carlo A, Bocci E, Taglieri L, Del Zotto L, Gallifuoco A. Energy analysis of an integrated plant: Fluidized bed steam gasification of hydrothermally treated biomass coupled to solid oxide fuel cells. Energies. 2021; **14**:7331

[31] Mutlu ÖÇ, Zeng T. Challenges and opportunities of modeling biomass gasification in Aspen Plus: A review.

Chemical Engineering & Technology. 2020;**43**:1674-1689

[32] Anukam A, Mamphweli S, Reddy P, Okoh O, Meyer E. An investigation into the impact of reaction temperature on various parameters during torrefaction of sugarcane bagasse relevant to gasification. Journal of Chemistry. 2015: 1-12

[33] Marcantonio V, Bocci E, Monarca D. Development of a chemical quasi-equilibrium model of biomass waste gasification in a fluidized-bed reactor by using Aspen Plus. Energies. 2020;**13**:53

[34] Safarian S, Unnþórsson R, Richter C. A review of biomass gasification modelling. Renewable and Sustainable Energy Reviews. 2019;**110**:378-391

[35] Kolb T, Aigner M, Kneer R, Mueller M, Weber R, Djordjevic N. Tackling the challenges in modelling entrained-flow gasification of low-grade feedstock. Journal of the Energy Institute. 2016;**89**:485-503

Chapter 6

Green and Sustainable Chemical Looping Plasma Process for Ammonia and Hydrogen Production

Mohsen Sarafraz, Farid Christo and Bernard Rolfe

Abstract

The overarching aim of this chapter is to propose a novel clean thermochemical process that harnesses thermal plasma technology to co-produce hydrogen and ammonia using a chemical looping process. The thermodynamic potential and feasibility of the process were demonstrated using a simulation of the system with aluminium and aluminium oxide as the oxygen and nitrogen carriers between the reactors. The effect of different operating parameters, such as feed ratio and temperature of the reactor, on the energetic performance of the process was investigated. Results showed that the nitridation and ammoniation reactors could operate at <1000 K, while the thermal plasma reactor could operate at much higher temperatures such as (> 6273 K) to reduce the alumina oxide to aluminium. The ratio of steam to aluminium nitride was identified as the key operating parameter for controlling the ammoniation reactor. Using a heat recovery unit, the extracted heat from the products was utilised to generate auxiliary steam for a combined cycle aiming at generating electricity for a thermal plasma reactor. It was demonstrated that the process can operate at an approximate self-sustaining factor ~ 0.11, and an exergy partitioning fraction of up to 0.65. Integrating the process with solar photovoltaic showed a solar share of ~32% without considering any battery storage units.

Keywords: three-stage chemical looping, ammonia, hydrogen, aluminium oxide, thermal plasma

1. Introduction

Ammonia is a key hydrogen carrier with the potential to be utilised as a fuel, an intermediate product, and also a combo product (of hydrogen and nitrogen) for agricultural and pharmaceutical applications. The current chemical pathway for generating ammonia is through the Haber-Bosch (HB) methane reforming process [1, 2] which is highly dependent on the quality and price of natural gas. The Haber-Bosch process includes consecutive stages of methane reforming, followed by water-gas shift improvement and catalytic nitrogen fixation at high pressure and low temperatures

[3, 4]. All these stages are energy-intensive, far from equilibrium and release CO_2 as part of the process due to the number of reformers existing in the process [5]. Hence, the HB process is not a sustainable cycle for nitrogen fixation [6] and ammonia production considering the technical challenges associated with environmental pollution and global warming [7].

The main issue with CO_2 formation is the presence and adjacency of oxygen and carbon that can generate either CO_2 via complete combustion, or CO through partial oxidation reactions as represented below:

$$C(s) + O_2(g) \rightarrow CO_2(g), \Delta G = -394.3 \text{ kJ/mol@273 K}, \tag{1}$$

and

$$2C(s) + O_2(g) \rightarrow 2CO(g), \Delta G = -269.9 \text{ kJ/mol@273 K}. \tag{2}$$

Reactions 1 and 2 can proceed depending on the availability of the oxygen in the reactor, referred to as the "O/C" molar ratio. In conventional combustors and gaseous reactors, these reactions will always occur due to the direct contact of oxygen and carbon. Therefore, new combustion systems should be developed to avoid the production of Green House Gases (GHGs). Chemical looping technology [8] is a promising thermochemical pathway by which the direct contact between air and fuel is inherently hindered, thereby avoiding the production of CO_2 and/or CO. In a chemical looping system, oxygen is transported between two reactors via an oxygen carrier (OC), which has the following properties:

- An oxygen carrier should provide sufficient oxygen content and reduction–oxidation capacity (cyclic redox) [9–11]. This means that the oxygen carrier should be able to release the oxygen where required and absorb it using a cyclic process for an unlimited time while retaining its solid physical properties, morphology, and structure [11, 12].

- The oxygen carrier should have suitable thermophysical properties such as high thermal conductivity, high heat capacity and density, hence representing favourable thermal diffusivity, $(\frac{k}{\rho c})$ [13]. It is a key thermal characteristic that affects the residence time of the oxygen carrier in the reduction and/or oxidation reactors. It will also promote thermal performance and heat transfer coefficient in the reactor [14].

- The oxygen carrier should also offer various stable states of oxidation [15]. Accordingly, depending on the operating conditions, end-user requirements and products, the chemical pathway can be designed based on these stable oxidation states.

A general chemical looping system includes reactors (for combustion [16], reforming [17], or gasification [18]) or as will be investigated here, it has three reactors (for ammonia/hydrogen production) [19] in which cyclic reduction and oxidation occurs.

There has been extensive research conducted on the potential of chemical looping systems for hydrogen production. For example, steam chemical looping gasification [20] or Co_2-chemical looping gasification [21, 22] is one potential configuration that has been extensively investigated for H_2-enrich fuel production. For example, He et al. [23] recently, proposed a new concept for steam chemical looping gasification to decrease the exergy destruction and to improve the cold gas efficiency of the gasifier. They used coal as the source of fuel and demonstrated that the efficiency of the cold gas can be as high as 86% in a three-stage chemical looping system. Also, the energy efficiency of 62.3% was calculated for the proposed plant. Pan et al. [24] used industrial waste such as gypsum and steel slag as the oxygen carrier to operate a steam chemical looping gasification system for generating H_2-enriched fuel. Based on the FactSage calculation [25] and experimental validation, they found that the optimum temperature for the gasification is around 1023 K, with a feedstock/oxygen carrier ratio of 1 and steam to oxygen carrier of 0.6. In a similar study, Zhao et al. [26] assessed a novel chemical looping gasification concept in which $NiFe_2O_4$ and $CuFe_2O_4$ particles were synthesised and utilised as an oxygen carrier for hydrogen production. They identified the optimum operating conditions for both reduction and oxidation reactions and showed that the gas yield can be as high as 64.98 mol/kg and syngas quality can reach 2.79 which is suitable for gas to liquid and Fischer-Tropsch processes [27, 28].

CO_2-chemical looping gasification is another process similar to dry gasification or pyrolysis which is used to generate syngas from coal or biomass. Recently, Xu et al. [22] developed thermodynamic models followed by conducting a series of experiments to assess the gasification of rice husk char sourced from pellet pyrolysis. They investigated the effect of various operating parameters including temperature, CO_2 concentration, and quantity of the oxygen carrier on the performance of the system. They showed that the presence of CO_2 can influence the performance of the gasification process. Also, they identified a mass transfer resistance within the char particles due to the change in the pore structure of the feedstock, which affected the gasification conversion. Zhang et al. [29] evaluated the effect of the red mud as the active oxygen carrier on the pyrolysis and gasification in a chemical looping gasification process. They utilised a fluidised bed reactor and investigated the effect of temperature (750–900°C) and also the ratio of the oxygen carrier to fuel (0.1–0.7) onthe syngas quality and chemical conversion extent of the fuel. The results indicate that red mud promoted the pyrolysis reaction and also intensified the gasification reaction. However, increasing the temperature decreased the syngas quality from 7.26 to 4.83. In another study, Zhang et al. investigated a novel hydrogen production process with integration of CaO - $Ca(OH)_2$ thermal storage unit for gasification of coal and biomass using chemical looping of calcium. The thermal storage unit decreased the quantity of oxygen carrier required for the process and also improved the rate of hydrogen production. at optimum conditions of temperature and pressure, the gas yield was 17% higher than a normal process. The process was reported to have an energy conversion efficiency of 42.1% together with an exergy efficiency of 39.4%.

Figure 1 represents schematic diagrams of two-stage and three-stage chemical looping technologies. In a three-stage chemical looping, it is assumed that the third reactor can dissociate the metal oxide into reduced metal directly and without any side reaction with air or other chemicals. However, in the two-stage chemical looping, the second reactor requires air to regenerate the oxygen content of the oxygen carrier.

As represented in **Figure 1a**, for combustion/gasification chemical looping systems, a reduction reaction occurs, in the fuel reactor between a metal oxide and a fuel, which is an endothermic reaction requiring an external energy resource to drive the reaction to its equilibrium. For example, for alumina oxide as an oxygen carrier in a

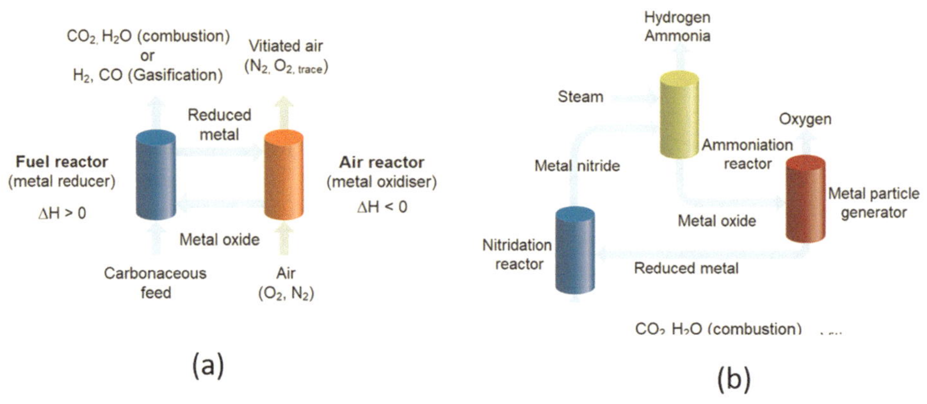

Figure 1.
Chemical looping systems for a) two-stage reforming/combustion/gasification and b) three-stage ammonia/hydrogen production.

chemical looping reforming system for CH_4, in the fuel reactor, methane reacts with aluminium oxide according to the following reaction:

$$3CH_4(g) + Al_2O_3(s) \rightarrow 2Al(s) + 6H_2(g) + 3CO(g), \Delta G_{1773\text{K}} = 5.1\ \text{kJ/mol} \tag{3}$$

or

$$1.5CH_4(g) + Al_2O_3(s) \rightarrow 2Al(s) + 3H_2(g) + 1.5CO_2(g), \Delta G_{1773\ \text{K}} = 5\ \text{kJ/mol} \tag{4}$$

The Gibbs value (ΔG) of both reactions are relatively the same. Hence, to determine the dominant reaction in the fuel reactor, the thermodynamic condition of the reactor must be studied, which includes the operating temperature and pressure of the fuel reactor, and the quantity of Al/aluminium oxide (oxygen availability, i.e., the O/C ratio). In the air reactor, the reduced metal (Al for our reactor) reacts with an abundant amount of air via an exothermic reaction to generate aluminium oxide particles and vitiated air. The vitiated air can then be utilised for electricity generation via gas combined power cycles, while the re-generated metal oxides can be fed into the fuel reactor again to be reduced over reforming reaction. As shown in the literature review section, it was identified that chemical looping gasification (either with steam or CO_2) is mainly used for generating syngas fuel. Accordingly, gasification includes the emission of CO_2 and CO as the products of the process. However, in the present work, the chemical looping technology will be utilised to generate hydrogen and ammonia instead of hydrogen and carbon monoxide, thereby either reducing or eliminating the carbon footprint from the process.

With process modifications, as represented in **Figure 1b**, the chemical looping technology can be utilised for the co-production of ammonia and hydrogen referred to as "3-stage chemical looping ammonia production, 3CLAP". In the nitridation reactor, a metal nitride can be formed by a direct exothermic reaction between a metal and a pure nitrogen stream. Pure nitrogen steam can be sourced from an air separation unit or any vitiated air from other industrial processes that have 99% nitrogen purity. In the ammoniation reaction, the metal nitride reacts with steam to generate ammonia and/or hydrogen depending on the availability of the steam, temperature of the reaction, and quantity of metal nitride in the reactor. The metal oxide produced from the ammoniation reactor is then fed into a dissociation reactor to be dissociated into pure metal and oxygen. While metal is fed into the nitridation reactor, oxygen can be chilled, stored and used for other industrial/health applications.

The main bottleneck of the 3CLAP system is the dissociation reactor. This is because dissociating metal oxides, such as Al_2O_3, requires high temperatures ($\sim$6273 K) to provide sufficient thermal driving force for such a highly positive Gibbs free energy of the "metal oxide to reduced metal reaction". Such high temperatures cannot be maintained by conventional reactors; hence, thermal plasma is suggested as a potential technology that can be utilised in this process.

In this chapter, the use of thermal plasma technology as a disruptive method is proposed to push the boundaries of dissociation reactors, and increase the chemical efficiency and performance of the process. Also, a renewable energy-friendly concept has been designed and simulated to demonstrate the feasibility of co-production of hydrogen and ammonia when integrated with photovoltaic solar energy generation. For the proposed process, Al/Al_2O_3 pair was used as it has reasonably large Gibbs free energy and a tendency to participate in the reaction, favourable thermophysical properties and stable thermodynamic phases during cyclic reduction and oxidation reactions. The Gibbs minimisation method coupled with thermochemical analysis was utilised to quantify the thermodynamic performance of the proposed system including hydrogen/ammonia ratio, exergy partitioned in the gas products, 1st law thermodynamic efficiency, self-sustaining factor, exergy partitioning factor, and renewable energy share (solar share).

2. Conceptual design of 3CLAP and its configuration

2.1 Process design

Figure 2 represents a detailed illustration of a three-stage chemical looping ammonia production system [19] designed for integration with a photovoltaic solar farm. In this process, three reactors for nitridation, ammoniation and thermal plasma

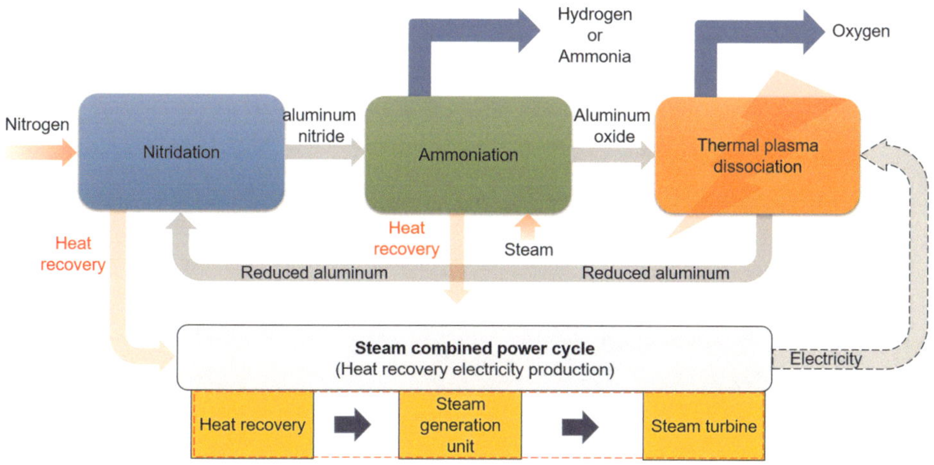

Figure 2.
An illustration of the proposed process for hydrogen and ammonia production using the aluminium chemical looping process.

are utilised amongst which thermal plasma is integrated with solar PV, while a heat recovery system is designed to recover heat from the reactors' outlets.

As already discussed, the energy of the thermal plasma is proposed to be supplied from renewable energy sources, such as solar photovoltaic combined with the electricity that can be generated from a built-in power cycle (steam combined cycle from the recovered heat of the processes), thereby improving the process sustainability and its energetic performance.

Figure 3 shows the process model developed using the Aspen Plus software package for simulating the 3CLAP process operating with alumina. Alumina was selected because it has high thermal conductivity, good thermal response, and high Gibbs free energy. Also, in the model, the aluminium/aluminium oxide pair was assumed to remain solid in all process stages. Also, the separators were assumed to have 100% separation efficiency and a capacity to separate solid materials from gases. The steam flow rate in the steam power cycle was varied depending on the thermal loading of the heat recovery thereby enabling the steam turbine to operate at a constant stream inlet temperature.

The simulations were performed with the Aspen Plus software package coupled with Matlab and Homer [30–32] to simulate the photovoltaic panels. To simulate the reactors, the Gibbs minimisation method [33] was utilised for the nitridation, ammoniation, and thermal plasma reactors. Also, thermochemical equilibrium analysis [34] was used to conduct sensitivity analysis of the reactors against operating parameters such as temperature or feed ratios. All Gibbs's free energy parameters, enthalpy of reactions, and thermodynamic properties were extracted from Barin's Handbook [35]. In addition, the following assumptions were considered in the model:

1. No heat loss occurs from any reactors during the process which represents an ideal operating condition;

2. All reactions proceed to equilibrium as there is no impurity in the system;

3. As the temperature is below the melting point, no structural change and morphology change occurs in the solid phase;

4. There is a robust particle handling system available to feed the particles in and from the reactors;

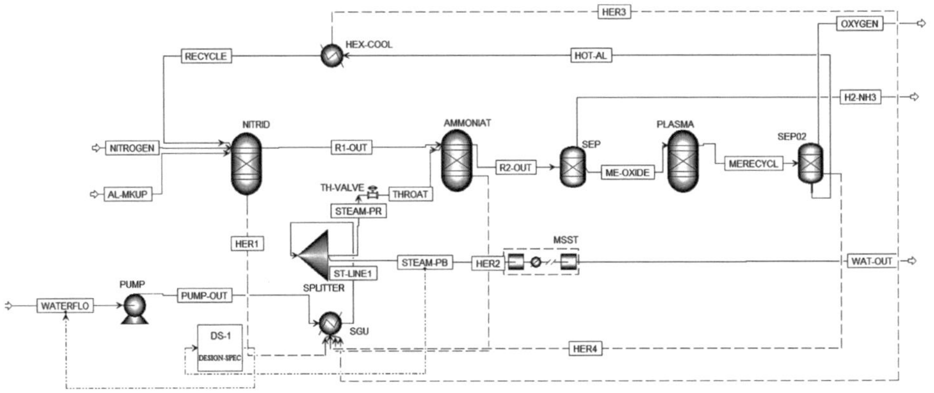

Figure 3.
The Aspen plus model was developed for simulating the 3CLAP system.

5. There are no side reactions between the walls of the reactors, and the solid particles;

6. There is a rapid quenching system for the thermal plasma reactor to cool the particles once they are formed inside the reactor to avoid driving other side reactions.

These assumptions are used to simplify the analysis and provide a benchmark of an ideal process, However, in practice, an adjustment based on the specific design is required in the model to account for the effect of imperfections and non-ideal conditions on the physical performance of the system.

2.2 Aluminium-oxygen-nitrogen thermodynamic system

As a preliminary step, the change in the Gibbs free energy of each reaction must be evaluated at equilibrium conditions to ensure reactions are thermodynamically feasible in fuel or air reactors. Accordingly, to assess the thermodynamics of the 3CLAP system at equilibrium conditions, it is necessary to consider aluminium, nitrogen and oxygen and any oxidation states such as Al_2O_3 as part of the thermodynamic system. Since all reactors work at atmospheric pressure, the Gibbs minimisation analysis is assumed to be only a function of temperature. In the nitridation reactor, the following reaction occurs:

$$Al(s) + N_2(g) \rightarrow AlN(s)\Delta G = 579.1\ \text{kJ/mol} \tag{5}$$

The AlN formed in the reactor is then fed into the ammoniation reactor in which the following reaction is expected to occur:

$$2AlN(s) + 3H_2O(g) \rightarrow Al_2O_3(s) + 2NH3(g), \Delta G_{298\ \text{K}} = 59.6\ \text{kJ/mol} \tag{6}$$

The key operating parameter here is the molar ratio of the reactants (steam/AlN ratio) which determines the final product of the reactor. Hence, depending on the availability of nitrogen and steam in the reactor, the following reaction can occur in the ammoniation reactor:

$$2AlN(s) + 3H_2O(g) \rightarrow Al_2O_3(s) + 3H_2(g) + N_2(g), \Delta G_{298\ \text{K}} = 321.8\ \text{kJ/mol} \tag{7}$$

The produced Al_2O_3 particles from both reactions are then fed into the thermal plasma reactor to be dissociated to aluminium and oxygen according to the following reaction:

$$Al_2O_3(s) \rightarrow 2Al(s) + 1.5O_2(g)\Delta G_{6373\text{K}} = -16.4\ \text{kJ/mol} \tag{8}$$

The pure aluminium is then returned to the nitridation reactor, completing the cycle, while oxygen can be chilled and stored for industrial and/or hygienic applications. This reaction is endothermic, thereby high energy is required in the thermal plasma reactor. Also, there is a need to utilise a rapid quenching system to be able to create aluminium particles in a plasma environment. The rapid quenching system must be robust, with a small response time close to that of a plasma reactor (i.e., microseconds to milliseconds). Considering the existing utilisation of plasma reactors for carbon black particle production [36], and the maximum temperature that a plasma reactor can reach [37], an arc thermal plasma is proposed. Combined with electricity produced via renewable energy sources, it decreases the emission of CO_2 and carbon footprint to the environment. As shown in **Figure 4**, a comparison between the equilibrium mole percentages of gaseous products in the thermal plasma reactor before and after the quenching shows that a rapid quenching system can operate at T > 6373 K to avoid recombination of oxygen and alumina. As shown in **Figure 4b**, the stable thermodynamic components after rapid quenching are Al_2O_3 and Al, while before quenching, O(g), O_2(g) and various solid phases of alumina are the dominant products of the thermal plasma.

2.3 Calculation of the thermodynamic parameter

To calculate the Gibbs free energy of the reactions and the change in the enthalpy of the proposed reactions, the following equation was utilised:

$$\Delta G_{R_i} = \sum \Delta G_{pro.} - \sum \Delta G_{Rct.}. \tag{9}$$

In this equation, G is the Gibbs free energy (kJ/mol), *pro* and Rct are acronyms for the products and reactants. If ΔG_{R_i}>0, it shows that the reaction is not thermodynamically feasible, while $\Delta G_{R_i} < 0$ shows that the reaction will proceed towards equilibrium without any thermodynamic barrier.

The same equation was utilised for the enthalpy of reaction, which is as follows:

$$\Delta H_{R_i} = \sum \Delta H_{pro.} - \sum \Delta H_{Rct.}. \tag{10}$$

In this equation, H is the enthalpy of reaction (kJ/mol). $\Delta H_{R_i}> 0$ shows that the reaction is endothermic thereby requiring thermal energy to proceed to the equilibrium state. However, $\Delta H_{R_i} < 0$ simply shows that reaction is exothermic, thereby releasing thermal energy during the reaction.

The self-sustaining parameter is defined as the ratio of the energy recovered and produced by the power block to the energy demand from the thermal plasma reactor defined with the following equation:

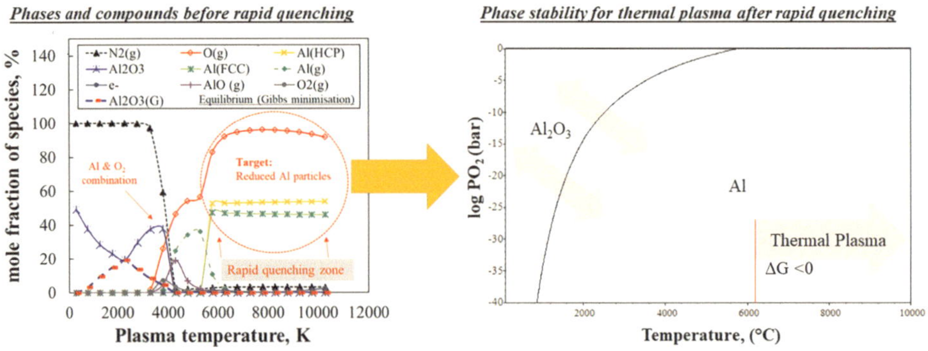

Figure 4.
Formation and stability of different phases and compounds in the thermal plasma reactor before and after the rapid quenching process for metals and powders [38].

$$SSF = \frac{W_{Tur.}}{E_{TPR}}. \tag{11}$$

In the above equation, $W_{Tur.}$ is the amount of energy recovered and converted to electricity using a multi-stage turbine from a built-in power block and E_{TPR} is the energy demand from a thermal plasma reactor.

The exergy partitioning factor is the amount of energy that can be stored in form of chemical exergy in the products of the plant and can be calculated using the following equation:

$$EPF = \frac{\dot{n}_{H2} LHV_{H2} + \dot{n}_{NH3} LHV_{NH3}}{\sum \Delta H_{net,reactors\,1,2,3}}. \tag{12}$$

Here, $\dot{n}_i$ is the mole flow of component I, LHV is the acronym defined for the lower heat value for either hydrogen or NH_3. The molar ratio of the steam to AlN is also defined as φ according to the following equation:

$$\varphi = \frac{[H2O]\left(\frac{kmol}{h}\right)}{[AlN]\left(\frac{kmol}{h}\right)}. \tag{13}$$

The above thermodynamic parameters were used in the calculations to assess the energetic performance and thermodynamic potential of the system.

3. Result and discussion

3.1 Nitridation reactor

Figure 5 represents the effect of temperature on the Gibbs free energy of the reaction in the nitridation reactor. As can be seen, by increasing the temperature of the reactor, for example from 250 K to 1000 K, the Gibbs free energy is decreased by ~25% recorded at atmospheric pressure. Also, the enthalpy of the reaction shows a

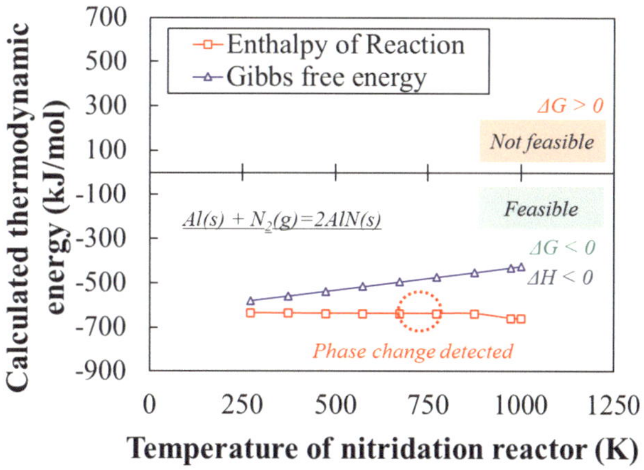

Figure 5.
Variation of the Gibbs free energy and enthalpy of the reaction with the temperature of the nitridation reactor.

linear weak dependence on the temperature of the reactor because it is an exothermic reaction. For example, increasing the temperature from 250 K to 1000 K causes only a 1.4% decrease in the enthalpy of the reaction. Also observed, was a slight drop in the enthalpy around T = 750 K due to the phase change in the AlN (solid) to the eutectic and liquid phase. Notably, to avoid complexity in the operation and reduce potential plant costs associated with maintaining the operating pressure, all reactors are designed at atmospheric pressure. This is advantageous as it can reduce erosion and corrosion related to high-pressure gaseous reactions in the reactors. It is worth mentioning that reaction 5 is spontaneous, hence it is not energy-intensive and does not require a continuous energy source to maintain its temperature. It can proceed to equilibrium due to the heat released during the reaction. Also, considering the melting temperature of the aluminium, the operating temperature of the nitridation reactor is set to the minimum with a 50 K buffer zone to avoid any liquid metals and clogs forming inside the pipes and walls of the nitridation reactor. Thus, T < 700 K is designated as a suitable temperature for the operation of the reactor. These findings are in good agreement with the results obtained in the literature for the hydrolysis reactors. Wang et al. [39] showed that a similar reaction can be driven with a non-equilibrium hydrolysis reactor in a system referred to as "Chemical Looping Ammonia Generation" (CLAG).

3.2 Ammoniation reactor

The generated AlN from the nitridation reactor is fed into the ammoniation reactor to be blended and react with steam for ammonia generation. As already elaborated in Section 3, reactions 6 and/or 7 occurs in this reactor depending on the operating conditions and molar ratio of steam to AlN feed. As represented in **Figure 6**, the results of the equilibrium analysis for the ammoniation reactor showed that while the dominant reaction in the reactor is reaction 6, it is an exothermic reaction. Also, there is a weak dependence of the performance of the reactor on temperature. Also, the Gibbs free energy for the reactions showed that with an increase in the temperature of the reactor, the ΔG value increases by ~30%.

For example, at T = 900-1000 K, −300 kJ/mol < ΔG < −400 kJ/mol. This shows that the reaction is highly spontaneous in the reactor at this operating temperature

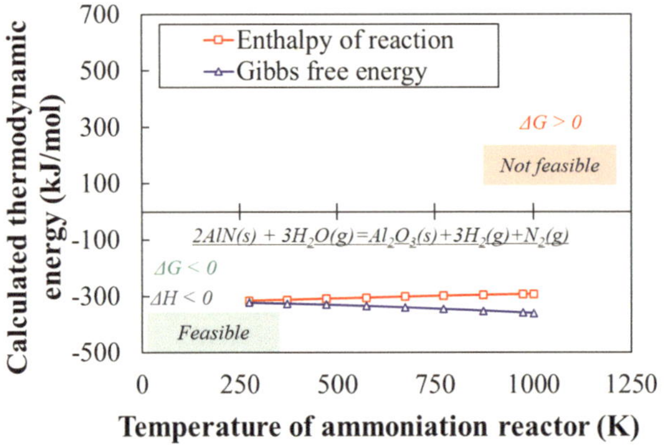

Figure 6.
The variation of the Gibbs free energy and enthalpy of reaction with the temperature of the ammoniation reactor.

region while at any other operating temperature < 1000 K, the reaction is still feasible and can reach equilibrium without any phase change occurring inside the reactor. Hence, the temperature of the reactor must be chosen wisely to avoid phase change and melting in the reactor. Overall, the upper limit operating temperature for the reactor is ~1000 K in which particles are still solid, the Gibbs free energy is negative and the reaction is exothermic.

As shown in **Figure 7**, the molar ratio of steam/AlN (φ value) is another key operating parameter that can control the chemical performance of the ammoniation reactor. Production rates of H_2 and NH_3 are nonlinearly dependent on the temperature of the ammoniation reactor, and it is proportional to the value of the steam to AlN ratio. For example, the final product from the ammoniation reactor could be determined by the φ value. As an example, at T = 540 K, for 0.5 < φ < 1, the dominant product of the reactor is hydrogen with H_2/NH_3 ratio ~ 25, while at 1 < φ <3, the H_2/NH_3 ratio significantly increases from 25 to 50 showing that production of ammonia is severely suppressed by increasing the amount of steam in the system. This is because, for larger φ values, the hydrogen and oxygen content in the reactor is increased which drives reaction 7 towards equilibrium. As a result, hydrogen production increases in the reactor. The best operating temperature range and φ values for producing hydrogen were at T > 500 K and φ = 3. While for ammonia production, not only the temperature of the reactor should be close to 300 K, but also the φ value should be below 0.5. The aluminium oxide produced in this reactor is then fed into the thermal plasma reactor to re-generate the reduced aluminium particles for completing the cycle and to provide the feed for the nitridation reactor. It is worth mentioning that in the hydrogen dominant operating region, still ammonia is produced, however, the mole fraction of the H_2: NH_3 is 1:300 or (less than 0.3%) that can be separated from the hydrogen stream using a robust aqua-ammonia condenser already developed in the literature [40] or hydrogen/ammonia membrane separators [41]. It is worth mentioning that alumina has already been identified as an oxygen carrier in a two-stage chemical looping ammonia production at 1773 K and 0.1 MPa using steam hydrolysis reaction. Galvez et al. [42] showed that both reactions can occur at atmospheric pressure without any added catalysts. They also advised that steam hydrolysis, which is an endothermic reaction, can be driven using concentrated solar thermal.

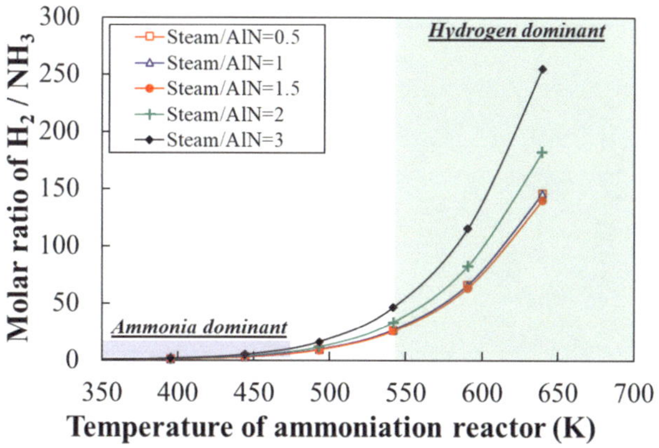

Figure 7.
Variation of the H_2/NH_3 ratio with φ values at various operating temperatures inside the ammoniation reactor.

3.3 Thermal plasma reactor

In a plasma reactive system, the reactions are driven by a non-conventional energy source. A plasma stream includes (but is not limited to) ions, charged particles, neutral species, radicals, and spinning-vibrational excited species with relatively high Gibbs free energy ($-\Delta G$) that participate in the reaction. The bombardment of any neutral species by electrons (electron avalanche) and charged species in plasma can stimulate the material to aggressively participate in the reaction. A thermal plasma is a type of reactive plasma configuration in which reactions occur at very high temperatures typically in the range 6000 K < T < 10,000 (for example for arc plasma reactor). The high temperature of the plasma reactor ensures that the reactions can proceed to equilibrium, thereby reaching a high chemical conversion extent and chemical efficiency (generally >99%). The results of the modelling for the Gibbs free energy and enthalpy of reaction, **Figure 8**, shows that at T > 6273 K, there is an inflexion point for the Gibbs free energy to a negative ΔG making the reaction thermodynamically feasible in the plasma reactor. For temperatures above 6273 K, an increase in the temperature of the thermal plasma reactor leads to an increase in the magnitude of the Gibbs free energy indicating that this reaction is more favourable at high-temperature conditions. Also, the enthalpy of the reaction is >0 (across the entire temperature range) implying the reaction is endothermic and requires thermal energy to be driven towards equilibrium. At such high temperatures, it is expected that all species exist in the gaseous phase thereby requiring a quenching system to generate solid particles from gaseous metals.

3.4 Effect of pressure

The proposed 3CLAP process at atmospheric pressure has several advantages:

1. there is no complexity associated with maintaining the pressure and temperature of each reactor. Since pressure and temperature can have a synergic or non-synergic effect on the performance of the reactors (e.g., nitridation and ammoniation), the operation of the reactors at atmospheric pressure can minimise the complexity and technical challenges due to the temperature–pressure interaction in the reactors.

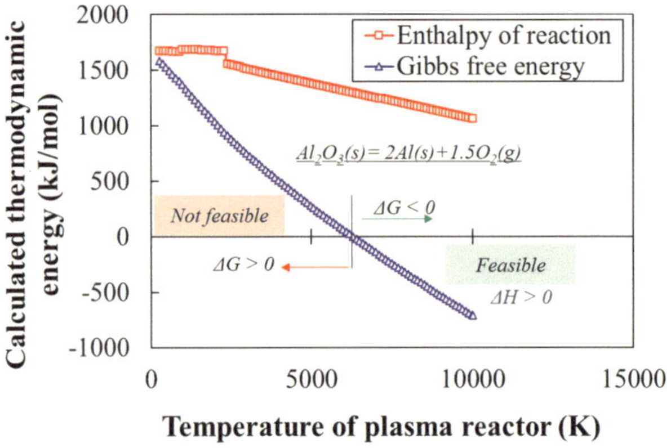

Figure 8.
Variation of the Gibbs free energy and enthalpy of reaction with the temperature inside the thermal plasma reactor.

2. Thermal plasma reactor [43] is a relatively new technology, and most reactors operate at atmospheric or slightly positive pressure conditions. Hence, it is technically challenging to pressurise a thermal plasma reactor that has internal delicate electrodes, narrow gaps, and thin walls.

3. Operations at high-pressure, high-temperature conditions [44, 45] create a harsh environment that requires specific design, construction materials (e.g., Inconel, fortified steel with passive protection, or carbon fibre). This strongly affects the economic viability of the process and weakens the justifications and techno-economic aspects of the plant. Hence, pressurising the reactors should be avoided as much as technically possible to reduce the Levelized Cost of Energy (LCOE) in the proposed system.

3.5 Thermodynamic characterisation of the process

The performance of the system based on the 1st law efficiency, nitrogen economy, steam consumption, exergy partitioning factor and self-sustaining factor are depicted in **Figure 9**. As can be seen, the 3CLAP process transport exergy in form of chemical exergy by partitioning it in hydrogen and ammonia as the main products of the system. Hence, a parameter is defined as Exergy Partitioning Factor to account for the fraction of energy that is partitioned in the ammonia and hydrogen. Additionally, using a heat recovery system, the energy recovered from outlet products and streams are recovered and converted into steam to produce electricity for supplying the energy demanded by the thermal plasma reactor. Hence, a parameter is defined as the Self-Sustaining Factor (SSF) defined as the ratio of electricity produced by the heat recovery block to the energy demand of the thermal plasma reactor. The results showed that the designed process can achieve an SSF value of ~0.11 without incorporating renewable energy or external energy resources.

Also, it was found that the nitrogen and steam economy of the process is >1.8 and < 1 showing that the proposed system can deliver 1 mol of ammonia at cost of

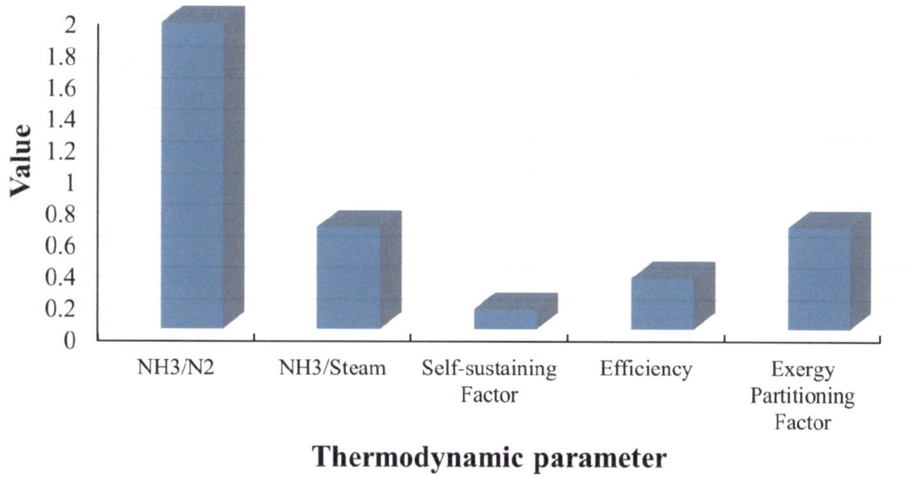

Figure 9.
The effect of the temperature of the thermal plasma reactor on self-sustaining and exergy partitioning factors.

consuming ~2 mol of nitrogen and < 1 mol of steam. Since steam production costs vary depending on the source of water, the proposed system can offer a plausible Levelized Cost of Energy and techno-economic value for ammonia production. Notably, to produce ammonia, the temperature of the ammoniation reactor should be as low as 300 K to promote the economic viability aspects of the ammoniation reactor.

3.6 Renewable energy penetration - integration with solar energy

In this section, an example of hybridisation of the 3CLAP with solar photovoltaic energy is represented to calculate the solar share and to obtain the installation capacity required for the PV panel. It further elaborates on which fraction of solar energy can penetrate the system, thereby reducing the CO_2 emission from the plant. While the entire thermodynamic plant is a zero-carbon process, energy demands for the reactors can contribute greenhouse gases to the environment. Thus, it is critical to demonstrate how the proposed process can operate with renewable energy. **Figure 10** represents the frequency of energy demand (**Figure 10a**) and its diurnal/annual fluctuations (**Figure 10b**) from the plant calculated for a case study to produce ~1 tonne/day of ammonia. The energy algorithm of the plant was defined such that using renewable energy and energy produced by the built-in power cycle are prioritised. However, if both sources cannot meet the demand of the process, the energy from the grid is ramped up to meet the demand.

A 2.1 MW generator was calculated for the grid network to adapt to demand fluctuations. Considering a chain of 4 kW DC to AC converters, the performance of the system for a location at Geelong, Victoria in Australia (where solar reception is moderate) was evaluated. As can be seen, the solar share increases with an increase in the PV installation capacity reaching 28.4% at PV installation capacity of ~5 MW and 33.6% at installation capacity of 10 MW. Notably, the solar share can be improved by adding battery storage units to the system to address the intermittent behaviour of solar irradiance. However, a detailed assessment of renewable energy penetration is not part of the main goal of this study.

4. Outlook and future of the technology

The proposed technology in this chapter offered the potential to co-produce ammonia and hydrogen via a built-in heat recovery system and photovoltaic solar energy.

The thermodynamic models developed in this chapter assessed the energetic performance of the process and identified the thermodynamic limitations associated with the use of a thermal plasma reactor. The optimum operating temperature, and feed conditions together with the phase stability diagram for the aluminium particles in the thermal plasma reactor were obtained via equilibrium chemical analysis based on the Gibbs minimisation method. The results of this study showed that:

1. From the thermodynamic aspect, the process is feasible, with plausible efficiency and exergy partitioning fraction, thereby, there is no thermodynamic barrier avoiding the operation of this cyclic process.

2. The thermodynamic potential of the system for integrating with renewable energy resources, such as photovoltaic solar energy, highlighted that the

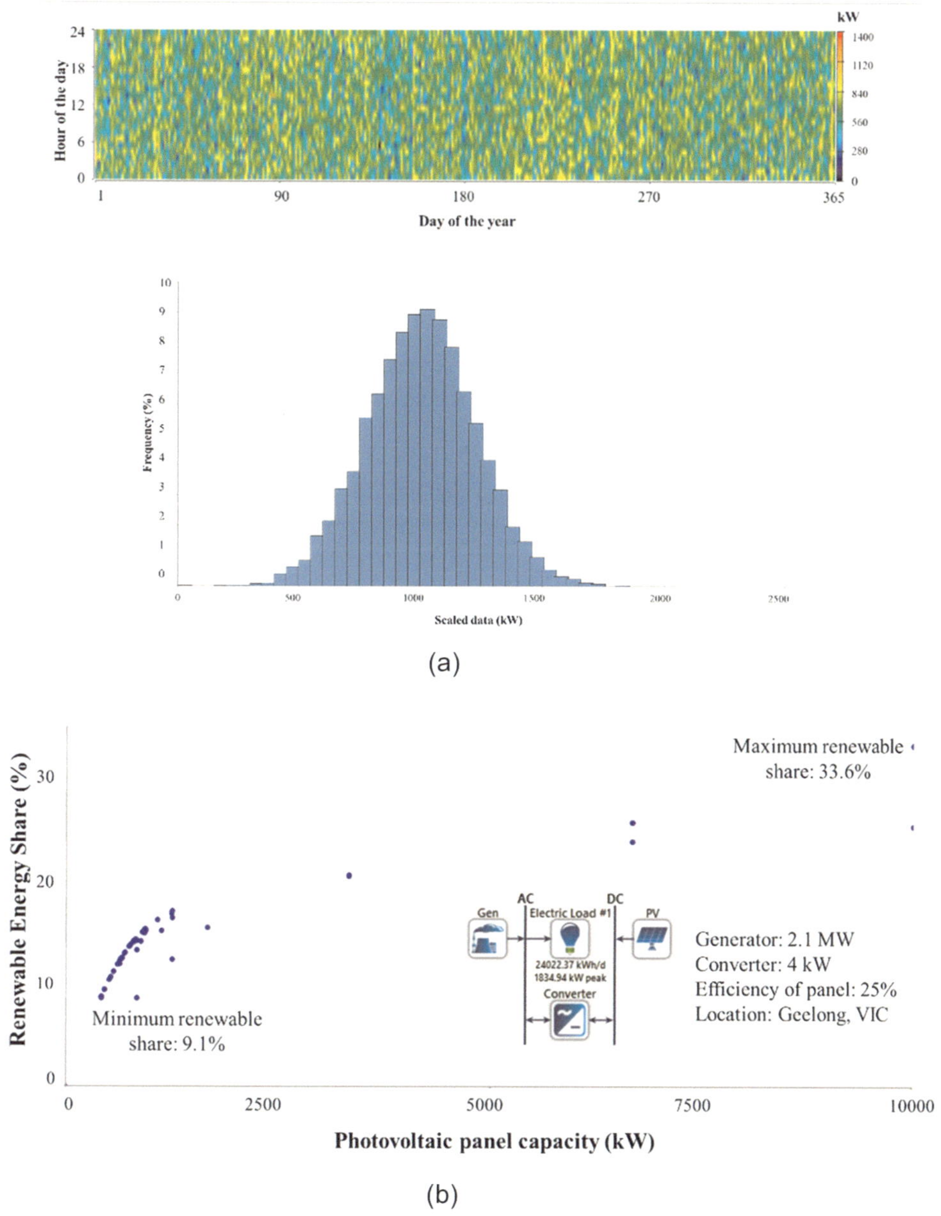

Figure 10.
(a) Scattering and distribution of the process plant dynamic energy demand data calculated with Aspen plus, (b) the calculated renewable energy fraction of the system based on the diurnal solar radiance profile obtained in Geelong, Victoria 3220.

proposed process is resilient when using renewable energy for either small-scale or large-scale production of ammonia and hydrogen. For example, at 1 tonne/day production, an instant solar share (share of renewable energy without any storage units) ranged from 21% to 33.6%. It means that by considering SSF = 10%, about 43% of the demand can be maintained by the renewable energy resource and heat recovery from the products.

3. The proposed process can have various industrial applications. For example, in a centralised mode, it can be integrated with an air separation unit (ASU). The ASU generates oxygen from the air and the side product of the plant is nitrogen (or vitiated air which has reduced oxygen). While oxygen is produced from both ASU and 3CLAP, the nitrogen from ASU can be fed in the proposed process to be converted to ammonia for fertiliser production. Also, the proposed process can be integrated with gas-cooled power plants (e.g., nuclear) not only to utilise the nitrogen from the cooling loop, but also to generate electricity for the thermal plasma and also to be injected into the grid. The other application of the proposed system is to be utilised in areas where solar reception is high. While a thermal plasma reactor can be built in any area with good solar reception, the rest of the plant can be developed next to water resources (for steam generation) and near nitrogen down streams. Hence, the proposed process can be designed in a decentralised (localised) mode.

While the feasibility of the process was successfully demonstrated in this chapter, further studies are still required to promote the technology readiness level of the 3CLAP. The required studies are listed as follows:

4.1 Rapid quenching technology

To handle gas–solid reactions in a thermal plasma reactor, a rapid quenching system is required to be fabricated to quench the aluminium particles once formed in the plasma reactor. The quenching process reduces the Gibbs's free energy and hinders the particles from participating in the unwanted and side reactions with the oxygen and radicals adjacent to the external surface of the aluminium particles. While the feasibility of the carbon black formation with thermal plasma has already been demonstrated in the literature, further studies are required to fortify the technology for solid particle formation. The rapid quenching process must have a small residence time and a very high heat transfer coefficient for high heat flux removal capacity.

4.2 Thermo-kinetics behaviour of the reactions

This study demonstrated the process feasibility of the 3CLAP process using equilibrium thermodynamics assuming all reactions were to proceed to the equilibrium state. The developed models are accurate for nitridation, ammoniation and for thermal plasma reactor (as it operates at high temperatures). However, there are possibilities that reactions are terminated at a point away from equilibrium. For example, the presence of impurities, localised temperature gradient, and changes in the morphology and structure of the aluminium particles can affect the cyclic behaviour of the process and suppress the chemical performance of the system. Therefore, there is a need to further study the thermo-kinetics behaviour of the aluminium particles and develop general kinetics models for the reactions in each reactor.

4.3 Material selection and design

Thermal plasma is a new disruptive technology, recently demonstrated in the lab-scale and pilot-scale studies [46]. Therefore, there is still a need to conduct further studies on the type of materials that can be used for constructing the reactor, electrodes, pipes and joints. While the temperature of the plasma reactor can locally reach

10,000 K [47], the bulk gas can reach a temperature up to 2000-3000 K. This requires a comprehensive study on material constraint and a material selection considering erosion, corrosion, and transient studies aiming at measuring tear and wear occurring inside the plasma reactor.

4.4 Prototyping the reactors

To the best of the authors' knowledge, this process is novel and there is no processing plant fabricated or constructed for ammonia/hydrogen production via thermal plasma plant. Hence, once kinetics studies are conducted, with the knowledge developed during material selection and design, prototypes can be constructed to measure the real-time efficiency, and performance of the 3CLAP process. In addition, there is further potential for other metals to be utilised in the process, thereby blending aluminium with other metals is another viable option to improve the real-life performance of the reactors.

5. Conclusions

In this book chapter, the potential of co-production of ammonia and hydrogen via a thermal plasma-assisted chemical looping technology was investigated. The proposed system was hybridised with solar energy (photovoltaic) and the renewable energy share was calculated. The Following conclusions were also made:

1. The Gibbs free energy calculations showed that all reactions in the nitridation, ammoniation and thermal plasma reactor are thermodynamically feasible with $\Delta G = 579.1$ kJ/mol, $\Delta G = 321.8$ kJ/mol and $\Delta G = -16.4$ kJ/mol, respectively. In the nitridation and ammoniation reactors, both reactions were spontaneous and both reactors could operate at $T < 1000$ K to exclude any phase change phenomenon inside the reactor.

2. A thermal plasma reactor was successfully demonstrated to dissociate Al_2O_3 to reduced aluminium particles at $T > 6273$ K with Al and Al_2O_3 as stable thermodynamic phases at $6273 \text{ K} < T < 10{,}000$ K. At $T < 6273$ K, $\Delta G > 0$ resulting causes reactions to be infeasible.

3. The Self-Sustaining Factor (SSF) and Exergy Partitioning Fraction (EPF) were calculated based on the energy demand of the thermal plasma and lower heating value (LHV) partitioned in the products, respective. It was identified that while the efficiency of the plant is >32%, the SSF and EPF are 0.11 and 0.65, respectively showing that the system can supply 11% of its energy using steam combined heat recovery and the exergy transported to the ammonia/hydrogen stream is 65%.

4. At a production capacity of 1 tonne/day (NH_3 basis), the integration of the 3CLAP with photovoltaic solar (with PV installation capacity of 3 MW up to 10 MW) showed that the instant renewable energy share can be as high as 21% to 33.6%, respectively (without using battery storage unit) which can be improved by adding battery storage or further integration with wind energy.

Overall, the thermodynamic assessments revealed that the 3CLAP system is considered as one of the processes for green ammonia and hydrogen production. However, its configuration can vary depending on the end-user requirements and geographical needs.

Acknowledgements

Dr. Mohsen Sarafraz gratefully acknowledges the financial support and research funding "PRESS 2022" received from Deakin University. Dr. Mohsen Sarafraz acknowledges the financial support received via the "Alfred Deakin Postdoctoral Research Fellowship".

Conflict of interest

The authors declare no conflict of interest.

Dedication

This book chapter is dedicated to the loving memory of my late mother, Fatemeh Sarraf, an angel whom I lost due to the COVID-19 outbreak in 2021. Also, it is dedicated to my beloved wife who supported me during the ups and downs of my life journey.

Author details

Mohsen Sarafraz[1]*, Farid Christo[2] and Bernard Rolfe[1]

1 School of Engineering, Deakin University, Geelong, Victoria, Australia

2 School of Aerospace Engineering and Aviation, RMIT University, Australia

*Address all correspondence to: mohsen.sarafraz@deakin.edu.au

References

[1] Kandemir T, Schuster ME, Senyshyn A, Behrens M, Schlögl R. The Haber–Bosch process revisited: On the real structure and stability of "ammonia iron" under working conditions. Angewandte Chemie International Edition. 2013;**52**:12723-12726

[2] Erisman JW, Sutton MA, Galloway J, Klimont Z, Winiwarter W. How a century of ammonia synthesis changed the world. Nature Geoscience. 2008;**1**: 636-639

[3] Leigh GJ. Haber-bosch and other industrial processes. In: Catalysts for Nitrogen Fixation. Dordrecht: Springer; 2004. pp. 33-54

[4] Moradi M, Ghorbani B, Ebrahimi A, Ziabasharhagh M. Process integration, energy and exergy analyses of a novel integrated system for cogeneration of liquid ammonia and power using liquefied natural gas regasification, CO2 capture unit and solar dish collectors. Journal of Environmental Chemical Engineering. 2021;**9**:106374

[5] Smith C, Hilla AK, Torrente-Murcianob L. Current and future role of Haber-Bosch ammonia in a carbon-free energy landscape. Energy & Environmental Science. 2020;**13**(2): 331-344

[6] Zheng J, Jiang L, Lyu Y, Jian SP, Wang S. Green synthesis of nitrogen-to-Ammonia fixation: Past, present, and future. Energy & Environmental Materials. 2021:1-6. DOI: 10.1002/eem2.12192

[7] MacFarlane DR, Cherepanov PV, Choi J, Suryanto BHR, Hodgetts RY, Bakker JM, et al. A roadmap to the ammonia economy. Joule. 2020;**4**: 1186-1205

[8] Luo M, Yi Y, Wang S, Wang Z, Du M, Pan J, et al. Review of hydrogen production using chemical-looping technology. Renewable and Sustainable Energy Reviews. 2018;**81**:3186-3214

[9] Kang K-S, Kim C-H, Bae K-K, Cho W-C, Kim S-H, Park C-S. Oxygen-carrier selection and thermal analysis of the chemical-looping process for hydrogen production. International Journal of Hydrogen Energy. 2010;**35**: 12246-12254

[10] Li F, Kim HR, Sridhar D, Wang F, Zeng L, Chen J, et al. Syngas chemical looping gasification process: Oxygen carrier particle selection and performance. Energy & Fuels. 2009;**23**: 4182-4189

[11] Adánez J, de Diego LF, García-Labiano F, Gayán P, Abad A, Palacios JM. Selection of oxygen carriers for chemical-looping combustion. Energy & Fuels. 2004;**18**:371-377

[12] Brown TA, Scala F, Scott SA, Dennis JS, Salatino P. The attrition behaviour of oxygen-carriers under inert and reacting conditions. Chemical Engineering Science. 2012;**71**:449-467

[13] Liu D-M, Tuan WH, Chiu C-C. Thermal diffusivity, heat capacity and thermal conductivity in Al2O3· Ni composite. Materials Science and Engineering: B. 1995;**31**:287-291

[14] Xing C, Li M, Fu Y, Chen X, Lu P, Li X, et al. Improving thermal diffusivity of supported Fe-based Fischer–Tropsch catalysts to enhance long-chain hydrocarbon production. Reaction Chemistry & Engineering. 2021;**6**: 1230-1237

[15] Ryu H-J, Kim J-W, Jo W-K, Park M-H. Selection of the best oxygen carrier particle for syngas Fueled chemical-looping combustor. Korean Chemical Engineering Research. 2007; **45**:506-514

[16] Abuelgasim S, Wang W, Abdalazeez A. A brief review for chemical looping combustion as a promising CO2 capture technology: Fundamentals and progress. Science of the Total Environment. 2021;**764**: 142892

[17] He Z, De Wilde J. Numerical simulation of commercial scale autothermal chemical looping reforming and bi-reforming for syngas production. Chemical Engineering Journal. 2021;**417**: 128088

[18] Nguyen NM, Alobaid F, Dieringer P, Epple B. Biomass-based chemical looping gasification: Overview and recent developments. Applied Sciences. 2021;**11**:7069

[19] Sarafraz MM, Christo FC. Sustainable three-stage chemical looping ammonia production (3CLAP) process. Energy Conversion and Management. 2021;**229**:113735

[20] Udomsirichakorn J, Salam PA. Review of hydrogen-enriched gas production from steam gasification of biomass: The prospect of CaO-based chemical looping gasification. Renewable and Sustainable Energy Reviews. 2014; **30**:565-579

[21] Sun H, Wang Z, Fang Y, Liu Z, Dong L, Zhou X, et al. A novel system of biomass for the generation of inherently separated syngas by combining chemical looping CO2-gasification and steam reforming process. Energy Conversion and Management. 2022;**251**:114876

[22] Xu J, Song T. CO2-gasification kinetics of biomass char with a red mud oxygen carrier for chemical looping combustion. Fuel. 2022;**313**:123011

[23] He S, Gao L, Dong R, Li S. A novel hydrogen production system based on the three-step coal gasification technology thermally coupled with the chemical looping combustion process. International Journal of Hydrogen Energy. 2022;**47**:7100-7112

[24] Pan Q, Ma L, Du W, Yang J, Ao R, Yin X, et al. Hydrogen-enriched syngas production by lignite chemical looping gasification with composite oxygen carriers of phosphogypsum and steel slag. Energy. 2022;**241**:122927

[25] Bale CW, Chartrand P, Degterov SA, Eriksson G, Hack K, Mahfoud RB, et al. FactSage thermochemical software and databases. Calphad. 2002;**26**: 189-228

[26] Zhao K, Fang X, Huang Z, Wei G, Zheng A, Zhao Z. Hydrogen-rich syngas production from chemical looping gasification of lignite by using NiFe2O4 and CuFe2O4 as oxygen carriers. Fuel. 2021;**303**:121269

[27] Dry ME. The fischer–tropsch process: 1950–2000. Catalysis Today. 2002;**71**:227-241

[28] Dry ME, Hoogendoorn JC. Technology of the Fischer-Tropsch process. Catalysis Reviews—Science and Engineering. 1981;**23**:265-278

[29] Zhang H-F, Chen L, Liu X-Y, Ge H-J, Song T, Shen L-H. Characteristics of cyanobacteria pyrolysis and gasification during chemical looping process with red mud oxygen carrier. Journal of Fuel

Chemistry and Technology. 2021;**49**: 1802-1810

[30] Givler T, Lilienthal P. Using HOMER Software, NREL's Micropower Optimization Model, to Explore the Role of Gen-Sets in Small Solar Power Systems; Case Study: Sri Lanka. Golden, CO (US): National Renewable Energy Lab; 2005

[31] Kangas P, Hannula I, Koukkari P, Hupa M. Modelling super-equilibrium in biomass gasification with the constrained Gibbs energy method. Fuel. 2014;**129**:86-94

[32] Koukkari P, Pajarre R, Hack K. Constrained Gibbs energy minimisation. International Journal of Materials Research. 2007;**98**:926-934

[33] Yahom A, Powell J, Pavarajarn V, Onbhuddha P, Charojrochkul S, Assabumrungrat S. Simulation and thermodynamic analysis of chemical looping reforming and CO2 enhanced chemical looping reforming. Chemical Engineering Research and Design. 2014; **92**:2575-2583

[34] Melgar A, Pérez JF, Laget H, Horillo A. Thermochemical equilibrium modelling of a gasifying process. Energy Conversion and Management. 2007;**48**: 59-67

[35] Barin I, Platzki G. Thermochemical Data of Pure Substances. Weinheim, VCh: Wiley Online Library; 1989

[36] Fulcheri L, Schwob Y. From methane to hydrogen, carbon black and water. International Journal of Hydrogen Energy. 1995;**20**:197-202

[37] Mostaghimi-Tehrani J, Pfender E. Effects of metallic vapor on the properties of an argon arc plasma. Plasma Chemistry and Plasma Processing. 1984;**4**:129-139

[38] Donaldson A, Cordes RA. Rapid plasma quenching for the production of ultrafine metal and ceramic powders. JOM. 2005;**57**:58-63

[39] Wang X, Su M, Zhao H. Process design and exergy cost analysis of a chemical looping ammonia generation system using AlN/Al2O3 as a nitrogen carrier. Energy. 2021;**230**:120767

[40] Kalogirou S. Recent patents in absorption cooling systems, recent patents on. Mechanical Engineering. 2008;**1**:58-64

[41] Maarefian M, Bandehali S, Azami S, Sanaeepur H, Moghadassi A. Hydrogen recovery from ammonia purge gas by a membrane separator: A simulation study. International Journal of Energy Research. 2019;**43**:8217-8229

[42] Gálvez ME, Halmann M, Steinfeld A. Ammonia production via a two-step Al2O3/AlN thermochemical cycle. 1. Thermodynamic, environmental, and economic analyses. Industrial & Engineering Chemistry Research. 2007; **46**:2042-2046

[43] Taylor PR, Pirzada SA. Thermal plasma processing of materials: A review. Advanced Performance Materials. 1994;**1**:35-50

[44] Min JK, Jeong JH, Ha MY, Kim KS. High temperature heat exchanger studies for applications to gas turbines. Heat and Mass Transfer. 2009;**46**:175-186

[45] Ho CK, Carlson M, Albrecht KJ, Ma Z, Jeter S, Nguyen CM. Evaluation of alternative designs for a high

temperature particle-to-SCO2 heat exchanger. American Society of Mechanical Engineers. 2019;**141**: V001T011A007

[46] Gomez E, Rani DA, Cheeseman CR, Deegan D, Wise M, Boccaccini AR. Thermal plasma technology for the treatment of wastes: A critical review. Journal of Hazardous Materials. 2009; **161**:614-626

[47] Mishin J, Vardelle M, Lesinski J, Fauchais P. Two-colour pyrometer for the statistical measurement of the surface temperature of particles under thermal plasma conditions. Journal of Physics E: Scientific Instruments. 1987; **20**:620

Chapter 7

Energy Efficiency Improvement in Surface Mining

Ali Soofastaei and Milad Fouladgar

Abstract

This chapter aims to provide an overview of energy efficiency in the mining industry with a particular focus on the role of fuel consumption in hauling operations in mining. Moreover, as the most costly aspect of surface mining with a significant environmental impact, diesel consumption will be investigated in this chapter. This research seeks to develop an advanced data analytics model to estimate the energy efficiency of haul trucks used in surface mines, with the ultimate goal of lowering operating costs. Predicting truck fuel consumption can be accomplished by first identifying the significant factors affecting fuel consumption: total resistance, truck payload, and truck speed. Second, developing a comprehensive analysis framework. This framework involves generating a fitness function from a model of the relationship between fuel consumption and its affecting factors. Third, the model is trained and tested using actual data from large surface mines in Australia, obtained through field research. Finally, an artificial neural network is selected to predict haul truck fuel consumption. The visualized results also clarify the general minimum areas in the plotted fuel consumption graphs. These areas potentially open a new window for researchers to develop optimization models to minimize haul truck fuel consumption in surface mines.

Keywords: energy efficiency, fuel consumption, surface mining, artificial intelligence, prediction

1. Introduction

Energy consumption in the last decade represents an increasing trend. Energy demand is growing across many countries globally as the population grows and human needs expand [1, 2]. As a consequence, fossil fuel consumption has also increased. Additionally, industrial activities have contributed directly and indirectly to annual greenhouse gas emissions [3, 4]. Australian energy consumption, for instance, grew by 0.7% a year on average for the past decade and reached 6014 PJ in 2019–2020, according to the Australian Bureau of Statistics [5, 6]. Fossil fuels (coal, oil, and gas) accounted for 93% of Australia's primary energy source in 2019–2020. Oil accounted for the most significant proportion of Australia's primary energy mix in 2019–2020, at 37%, followed by coal (28%), gas (27%), and renewable energy 8% [5].

Therefore, in recent years, the effects of sustainability on energy production and use have been well understood, and sustainability studies have recently considered enhancing energy efficiency. This is not an outlier trend to be found within the mining industry.

For many countries, mining is a crucial industry. Minerals, coal, metals, sand, and gravel are needed for construction and production and provide employment, taxes, and dividends that fund hospital, schools, and public facilities. To put it another way, mining is first and foremost a source of valuable mineral raw materials that are considered essential by all countries for national security, wealth creation, maintaining and improving the living standards of individual citizens [7].

Mining operations consume vast amounts of energy. For example, Mining in Australia consumes more than 9% of the nation's total energy consumption, which amounts to 570 petajoules per year [8]. Approximately 41% of mining's energy is derived from diesel, 33% from natural gas, and 22% from grid electricity, with the remainder being derived from coal, LPG, renewables, and biofuels. It is worth noting that diesel consumption has recently decreased from 49 to 41% in a decade [9]. It has been replaced mainly by natural gas and grid electricity due to infrastructure development and fluctuations in oil prices.

The mining industry appears to have benefited from rising fuel prices in the 1970s, as evidenced by studies on improving energy efficiency and using sustainable energy sources in the industry. As a result, reducing energy consumption has gradually become a priority for many countries with significant mining operations. Several projects have been conducted by the United States, Australia, Germany, Canada, and China that reduce energy consumption in mining operations [10–12]. Moreover, some governmental moves make industries pay for carbon taxes and similar regulatory costs, leading to the unprofitable and unsustainability of energy-intensive processes.

There are several aspects in the mining value chain where energy efficiency can be improved, such as managing electricity demand, capturing waste heat, improving ventilation, reducing mine drainage, and generating energy from by-products [13]. Numerous authors examined the energy consumption of various mining equipment. Oskouei and Awuah-Offei [11] studied energy consumption and dragline parameters. Peralta and colleagues demonstrate in their research that a maintenance policy based on equipment reliability can significantly reduce energy consumption [14]. Kuzin and colleagues proposed a method for estimating the energy consumption of process equipment and the relationship between energy consumption and vibration parameters and the temperature of the equipment surfaces [15]. According to research, blasting and material handling operations such as loading and hauling have the most significant potential for improving energy efficiency and lowering operating costs [16–20]. Based on numerous studies that have been conducted comparing energy efficiency improvements in mineral processing plants and material handling to other processes, this statement is confirmed.

Companies in the mining industry have recently begun implementing advanced Information Technologies (IT) to improve processes and simultaneously reduce energy consumption and operating costs. The mining industry deals with a large amount of data with layers of hidden knowledge. Since data analytics involves the science of analyzing raw data to derive information, it is a very effective technique for bringing disparate data sources together. Furthermore, data analytics provides cost savings, faster and better decision-making, and the development of new products and services, among other benefits [21]. As a result, data analytics is widely used, and it has a wide range of applications that many people may not have previously considered.

This chapter discusses advanced data analytics techniques to enhance mining energy efficiency. Open-pit haulage is the main target of the discussion. One of the objectives of this research is to develop a sophisticated data analytics model for assessing haul truck energy efficiency in surface mining. Concerning energy consumption in surface mining, the primary focus of this research is on the application of Artificial Neural Networks (ANNs) for prediction in the investigation of energy efficiency.

2. Energy consumption in mining

Mining is a crucial part of the global economy. In 2020, the top 40 mining companies made approximately 656 billion dollars [22]. Every year, hundreds of millions of raw materials are delivered to factories, the construction industry, utilities, and other commercial enterprises in the United States. Coal, metals, minerals, as well as sand and gravel are examples of such resources.

Research conducted in this area focuses on mining in Australia, which has also been a cornerstone of the Australian economy. Australia is the world's largest producer of lithium and is one of the world's top five producers of gold, iron ore, lead, zinc, and nickel, as well as some other minerals. In addition, the country has the most significant uranium and fourth-largest black coal resources in the world, respectively. Minerals are also one of Australia's major exports. Depending on their location, they are mined through open-cut mining on the earth's surface or underground mining techniques.

About the population, the energy consumption of Australia's industrial sector is among the highest. However, partly due to lower energy prices and lower rates of capital investment in the manufacturing industry, the rate of improvement in Australia's industrial development has lagged behind that of other countries [23, 24].

According to the most recent statistics, the sectors with the highest energy consumption in 2019–2020 were manufacturing and mining [8]. The mining industry in Australia consumes about 570 Pita Joules (PJ) of energy each year. However, Approximately a tenth of it can be savable [18]. Due to the significant energy savings opportunities, mining firms and the government have conducted many studies on cutting this industry's energy consumption (see **Table 1**).

The amount of energy consumed by a mine depends on various factors, including the minerals it mines, the production processes it employs, and the extraction technologies it employs. **Figure 1** illustrates the relative amounts of energy used by the world's three most energy-intensive mining sectors.

A mine's fuel type will vary depending on its type (underground or open-pit mine) and its process. Mining operations use diesel fuel, electricity, natural gas, coal, and gasoline, which account for 34%, 32%, 22%, 10%, and 2% of total energy consumption, respectively (see **Table 2**).

Table 3 shows how much energy is currently being used by various types of mining equipment. The most energy is used in material handling by diesel equipment (17%) and grinding equipment (40%).

Case study	2015–2016 (PJ)	2016–2017 (PJ)	2017–2018 (PJ)	2018–2019 (PJ)	2019–2020 (PJ)
Agriculture[a]	109	115	117	118	104
Mining	524	529	547	562	570
Manufacturing	964	928	938	915	910
Electricity[b]	136	130	132	130	133
Construction	135	142	148	140	144
Transport[c]	646	665	688	693	606
Commercial and services	330	337	345	354	347
Households	1247	1255	1274	1279	1228

[a]*Includes Forestry and fishing.*
[b]*Includes Gas, water supply, and waste services.*
[c]*Includes Postal and warehousing.*

Table 1.
Energy consumption by industries and households in Australia [8].

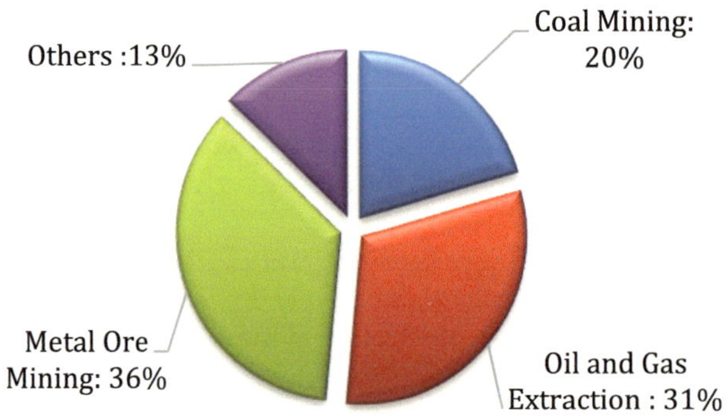

Figure 1.
Energy use by mining sub-division (PJ/year) [6, 25].

Fuel type	Amount (PJ/year)	Percentage
Gasoline	12.1	2%
Coal	60.7	10%
Natrural gas	133.5	22%
Electricity	194.2	32%
Diesel	206.4	34%

Table 2.
Fuel consumed in the mining industry [6, 26, 27].

Fuel type	Amount (PJ/Year)	Percentage
Electric Equipment for Material Handling	9	4%
Separations	9	4%
Ancillary Operations	19	8%
Crushing	9	4%
Ventilation	23	10%
Digging	14	6%
Blasting	5	2%
Drilling	12	5%
Grinding	93	40%
Material Handling Diesel Equipment	40	17%

Table 3.
Energy consumption in the mining industry [6, 26, 27].

3. Truck energy consumption

The hauling of mined material from a pit to a stockpile, dumpsite, or the next step in the mining process is accomplished by trucks at a surface mining operation. Their use may be combined with other types of machinery, such as loaders, diggers, and excavators, depending on the layout and production capacity of the site [28–31]. Surface mines in Australia use a considerable amount of diesel and are costly to purchase, maintain, and operate [28].

It is insufficient to analyze only the parameters specific to a haul truck to estimate its energy efficiency. By expanding the analysis of how energy is used throughout an entire fleet, companies can often find more significant benefits [32, 33]. This chapter is concerned with the identification and optimization of these parameters.

A fleet's energy efficiency can be affected by a variety of factors, including the rate of mining at a particular site, the age and condition of its equipment, the payload, the truck speed, and truck cycle time, the mine layout and plan, the idle time, tire wear, rolling resistance, dumpsite design, engine operating parameters, and shift patterns. By combining this knowledge with mine planning and design procedures, energy efficiency can be improved [34–38].

3.1 Mine operating parameters

Trucks in mines can use a variety of parameters that can influence how much energy they use, some of which are listed in **Table 4**.

3.2 Truck travel time

The time spent hauling and returning the payload is referred to as the travel time. There are four methods for calculating travel time: time study, Rimpull curves, empirical calculations, and computer simulation. Time study is the most common method.

Fuel type	Amount (PJ/year)
Truck model and type	Each type and model of the truck has unique characteristics, and these can be effective on energy consumption by truck.
Material	Material that is hauled
Bucket density	The density of the material being loaded.
Swell factor	The swell factor is the volume increase after material has been disturbed.
Bucket load	Estimated bucket load that the loading unit can carry in BCM
Calculated passes to fill	Estimate how many bucket loads (passes) are required to fill the truck to its nominal capacity.
Calculated truck payload	The estimated average payload that the truck will carry after considering all the above factors
Load factor	Percentage of truck fill compared to its nominal or rated payload.
Time per pass	Time is taken for a loading unit to complete one pass.
Load time	Time is taken to load the truck.
Spot time	The time during which the loading unit has the bucket in place to dump but is waiting for the truck to move into position. Spot time will depend on the truck driver's ability and the loading system. Double-side loading should almost eliminate spot time.
Dump time	Time is taken for the truck to maneuver and dump its load either at a crusher or dump.
Fixed time	Sum of load, spot, and dump time. It is called 'fixed' because it is essentially invariable for a truck and loading unit combination.
Travel time	Time is taken to haul and return the load.
Wait time	Duration of time spent waiting for the loading unit to arrive.
Cycle time	The truck's round trip time is the sum of fixed, travel, and wait times.
Efficiency	The amount of productive time achieved in one hour of operating time is measured. The following activities are included in the efficiency factor: Cleaning up by the loading unit or dozer and grading. All aspects to consider are slowdowns in the crusher and dump, fueling, inspections, loading unit movement, and operator experience. Under the heading of trucking, Weather-related delays have occurred more frequently than usual.
Queue factor	It keeps track of the time that has been lost due to queuing. It is yet another way of expressing the length of time spent waiting.
Productivity	Tonnes of production hauled in an operating hour (t/h) Productivity = Efficiency/(Cycle time × Truck payload × Queuing factor)
Mechanical availability	Depending on the type of machine, its age, and the maintenance philosophy,
Utilization	Operating time divided by available time
Production	Hourly Productivity × Operating Hours

Table 4.
Parameters that influence the energy consumption of haul trucks [21, 39, 40].

3.3 Haul profile

Information that effectively estimates travel time, such as distance, vehicle weight, slope, and speed limit, is called a haul profile.

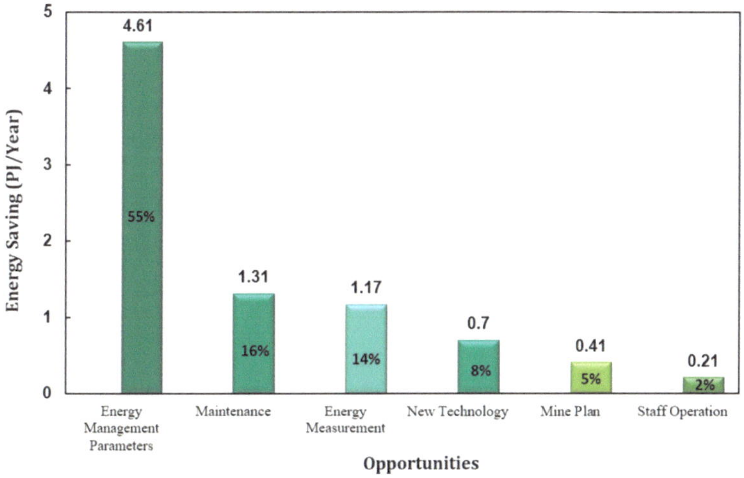

Figure 2.
Opportunities for energy conservation in the mining industry [41, 42].

4. Identifying the most influential parameters

A variety of variables influences the energy consumption of a truck. Because of the constraints of the project, it is not possible to model all of the parameters at this time. The model, therefore, includes the most important parameters. Mining energy savings opportunities can be categorized based on the latest government reports, staff operations, maintenance procedures, management systems, energy measurement, energy management parameters, and new technologies [41–43]. **Figure 2** represents the amount of energy saved and the percentage of total savings achieved by mining companies during the 2019–2020 period, based on the types of energy efficiency opportunities identified and implemented by the companies. The mining entities identified the most energy savings opportunities through energy management projects or 4.61 PJ. This accounts for 55 percent of the total potential savings determined by the mining companies.

Three main parameters have been identified as effective in reducing truck fuel consumption due to an online survey conducted for this research. The survey reached out to 60 industry professionals, who responded at a rate of 81 percent. According to the survey findings, the payload, truck speed, and the resistance of the road are the three most important factors influencing haul truck fuel consumption. Following identifying the primary effective parameters on haul truck fuel consumption in surface mines, a practical method for creating the model must be selected to predict the burnt fuel with the trucks in the mine site. ANN is the name of this method.

5. Artificial neural network (ANN)

ANNs or neural networks, also known as a simulated neural network (SNN), or what is known as 'parallel distributed processing,' represents how the brain uses various methods to learn. The ANN is a collection of mathematical models intended to mimic a few of the common characteristics of natural neural networks. In some cases,

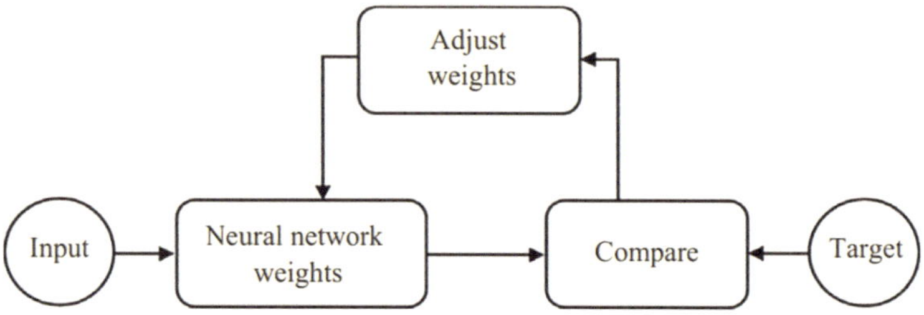

Figure 3.
An example of a typical artificial neural network procedure [44].

the unusual structure of the data processing system may be the most critical component of an ANN paradigm. **Figure 3** depicts an example of a neuronal model that consists of weighted connectors, an adder, and an activation function, among other components. These models are used in computer applications to solve complex problems that arise from user input. They do not require a mathematical description of the process-related phenomena, nor do they need any information to identify the factors that are associated with the process. Instead, they rely on acceptable errors and simple models [35, 36].

In neural networks, the node is the main component. Signals from various sources are summarized by biological nodes, which perform nonlinear operations on the results to produce output. When it comes to artificial neural networks, they are typically divided into three layers: an input layer, a hidden layer, and an output layer. According to its most basic configuration, each of the inputs and its associated weights is multiplied by the connected weight of its neighboring input. The resulting quantities and biases pass through activation functions to produce the output.

6. Proposed model

Several different variables influence fuel consumption for haul trucks. The performance of a typical haul truck is illustrated in **Figure 4** by the key factors that influence it.

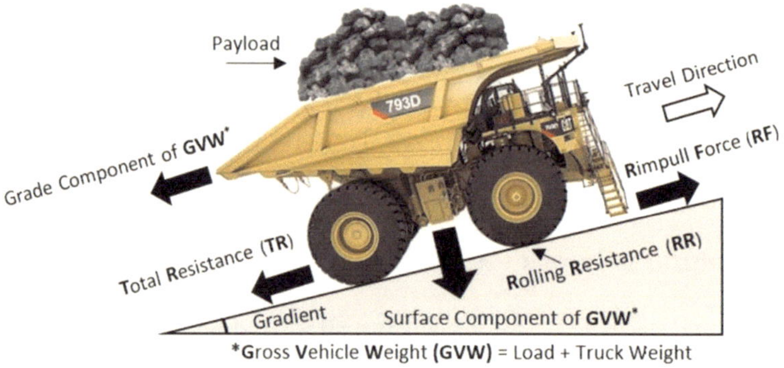

Figure 4.
Influential critical factors of performance of a typical haul truck.

The results of this study examined the effects of the Payload (L), Truck Speed (S), and Total Resistance (TR) on fuel consumption. Burt et al. define the TR as the sum of the Rolling Resistance (RR) and the Grade Resistance (GR) [45].

$$TR = RR + GR \tag{1}$$

When the characteristics of the tires and the haul roads are considered, this RR can be used to calculate the Rimpull Force (RF). As the truck tire rolls down the haul road, the RF measures the resistance to motion in the tire. The GR denotes the gradient of the haul road. When expressed in percentage, it is determined by the relationship between the rise of the road and the horizontal length. The truck's Fuel Consumption (FC) can be calculated with the help of Eq. (2) [46]:

Eq. (2) (Filas 2002) can be used to calculate FC.

$$FC = \frac{SFC}{FD}(LF.P) \tag{2}$$

Where SFC is the engine Specific Fuel Consumption at full power (0.213–0.268 kg/kW hr) and FD is the Fuel Density (0.85 kg/L for diesel). The simplified version of Eq. (2) is presented by Runge [47]:

$$FC = 0.3(LF.P) \tag{3}$$

LF is the engine Load Factor and is defined as the ratio of average load to the maximum load in an operating cycle [48], p is the truck power (kW), and it is determined by:

$$P = \frac{1}{3.6}(RF.S) \tag{4}$$

The calculation mentioned above method does not work ideally in mine sites. The calculated consumed fuel by haul trucks using the simple formula same as Eq. (3) cannot help mine managers, operation team, and other related groups estimate fuel consumption. The accuracy of proposed straightforward approaches by researchers is not enough to allow the mine managers to make the correct decisions and improve the energy efficiency in surface mines. Based on the reasons mentioned above and to solve the business problem, this chapter introduces an innovative solution using ANN to predict truck fuel consumption based on the collected data for three effective parameters: payload, truck speed, and total resistance.

6.1 Developed ANN model

Biological nodes generate outputs by combining signals from various sources nonlinearly. A neural network is typically composed of three layers: an input layer, one or more hidden layers, and an output layer, among other things. In its most basic form, each input is multiplied by the weight of the connected input, and the result is passed through the activation functions to generate the output (see Eqs. (5)–(7)).

$$E_K = \sum_{J=1}^{Q}\left(W_{I,J,K}X_J + B_{I,K}\right) \qquad K\text{-}1,\ 2,\ \ldots,\ M \tag{5}$$

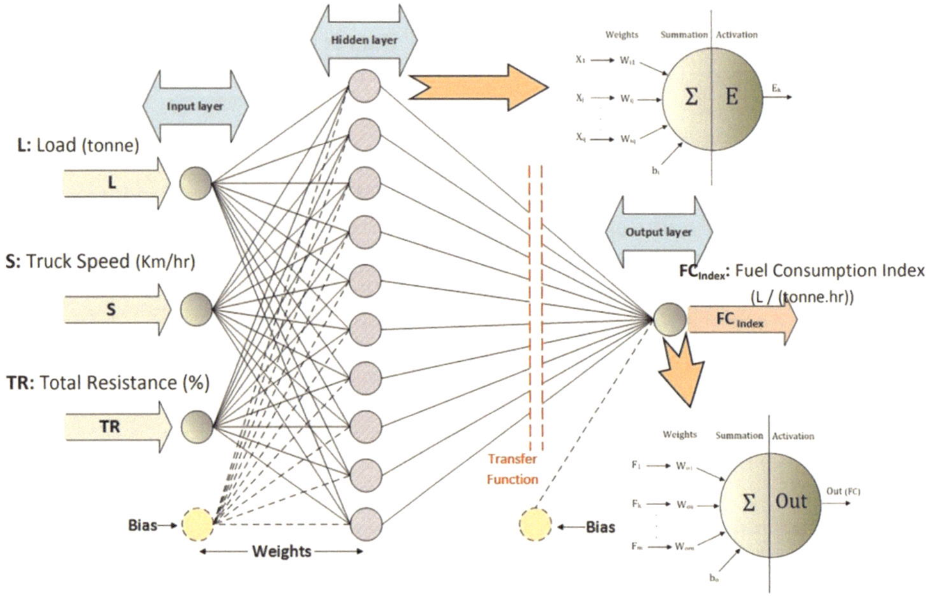

Figure 5.
Structure of ANN developed model.

Where x is the normalized input variable, w is the weight of that variable, i is the input, b is the bias, q is the number of input variables, and k and m are the counter and number of neural network nodes, respectively, in the hidden layer.

Figure 5 depicts a simplified representation of the structure of the model developed in this research. It should be noted that the hidden layer nodes are free to generate their output using any differentiable activation function they choose.

In general, the activation functions are made up of both linear and nonlinear equations, depending on the situation. Matrixes $W_{i,j,k}$, and $b_{i,k}$ are used to organize the coefficients associated with the hidden layer in the hidden layer. As an activation function between the hidden and output layers, Eq. (6) can be used to achieve the desired result (in this Equation, f is the transfer function).

$$F_K = F(E_K) \tag{6}$$

During the output layer's computation, the hidden layer's signals are weighted summed, and the coefficients associated with these weights are organized into three matrices: $W_{o,k}$, and B_o. The network's output can be calculated using matrix notation, as shown in Eq. (7).

$$OUT = \left(\sum_{K=1}^{M} W_{O,K} F_K \right) + B_O \tag{7}$$

It is presented in this chapter the results of a study in which different types of algorithms were investigated to determine the best back-propagation generating algorithm. First, let us compare the Levenberg-Marquardt (LM) back-propagation generating algorithm to other similar algorithms. It has the lowest mean square error (MSE), Root mean square error (RMSE), and Correlation Coefficient (R^2) of any of the algorithms (see Eqs. (8)–(10)). In addition, network generation using the LM algorithm can be accomplished with the smallest possible Expanded Memory Specification (EMS) and a quick generating process by using the LM algorithm. The statistical criteria MSE, RMSE, and R^2 are used to evaluate the accuracy of the results in accordance with the following Equations (Ohdar and Pasha 2003 and Poshal and Ganesan 2008), which are as follows:

$$\mathrm{MSE} = \frac{1}{p}\sum_{r=1}^{p}\left(y_r - z_r\right)^2 \tag{8}$$

$$\mathrm{RMSE} = \left(\frac{1}{p}\sum_{r=1}^{p}\left(y_r - z_r\right)^2\right)^{\frac{1}{2}} \tag{9}$$

$$R^2 = 1 - \frac{\sum_{r=1}^{p}\left(y_r - z_r\right)^2}{\sum_{r=1}^{p}\left(y_r - \bar{y}\right)^2} \tag{10}$$

Where y denotes the target (actual), z denotes the output (estimated) of the model, (y denotes the average value of the targets, and p denotes the number of network outputs). To examine the error and performance of the neural network output, the MSE and R^2 methods were used. In addition, the LM optimization algorithm was used to determine the optimal weights for the network.

The proposed ANN model for function approximation has the structure of a feed-forward multi-layer perceptron neural network with three input variables and a single output. One or more hidden layers of sigmoid nodes are frequently found in the feed-forward network, tracked by an output layer of linear nodes. Nodes with nonlinear activation functions are arranged in multiple layers, allowing the network to learn the linear and nonlinear connections between the input and output vectors over time. The linear output layer enables the network to generate values outside the [−1,+1] range using a linear function. The activation functions in the hidden layer (f) are the continuous differentiable nonlinear tangents sigmoid presented by Eq. (11).

$$f = \tan\ \mathrm{sig}(E) = \frac{2}{1 + \exp(-2E)} - 1 \tag{11}$$

When determining the optimal number of nodes in the hidden layer, MSE and R^2 were calculated for various hidden layer densities to determine their optimal number of nodes. For 15 nodes in the hidden layer, the minimum MSE and the maximum R^2 (best performance) were discovered, resulting in the best overall performance (as shown in **Figure 6**).

To train the ANN model, 4600 pairing data points were randomly selected from the 6630 values of the site data that had been gathered for this study (A large surface mine located in central Queensland, Australia). The values of payload, Vmax, and TR were calculated from the site data and used to train the ANN model, which was then used to calculate the fuel consumption from the site data.

As shown in **Figure 7**, the variation of MSE occurs during the network training process: it can be seen that the error approaches zero after 25 epochs, which indicates that the desired network convergence was achieved during the training process.

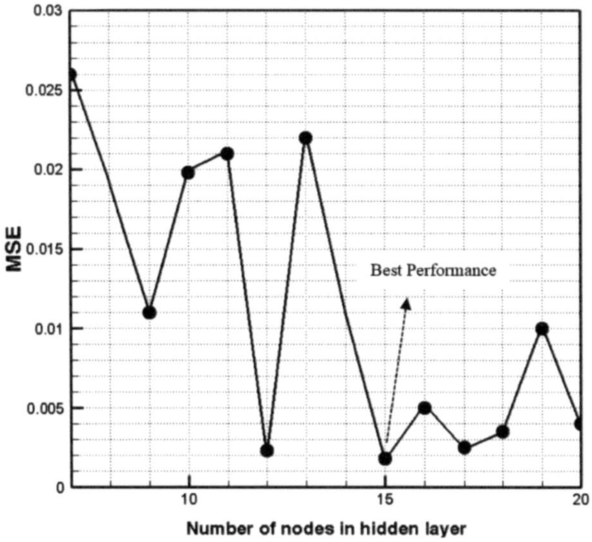

Figure 6.
The performance of the network at different hidden nodes using the LM algorithm.

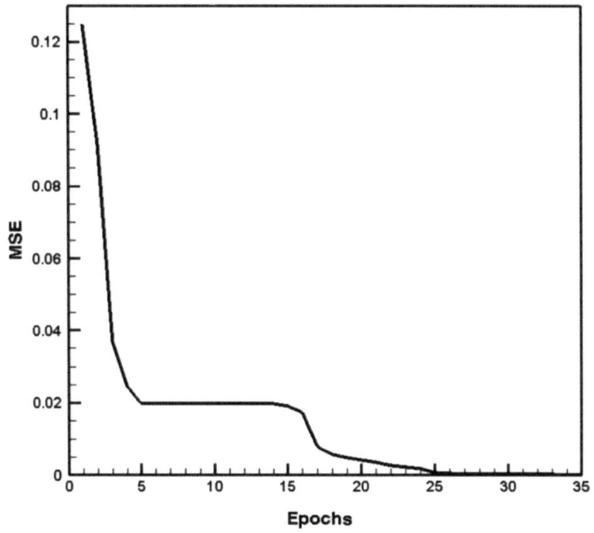

Figure 7.
Neural network error diagram (MSE) during network training.

Approximately 2030 independent samples were used to evaluate the accuracy of the network and validate the model. The test results of the synthesized network are depicted in **Figure 8**, where the vertical and horizontal axes represent the estimated fuel consumption values by the model and the actual fuel consumption values, respectively, and the vertical and horizontal axes represent the actual fuel consumption values.

Figure 8 illustrates the accuracy of the developed model. The results show more than 85% accuracy, which is acceptable for a mining application using unstructured noisy data collected from a real mime site.

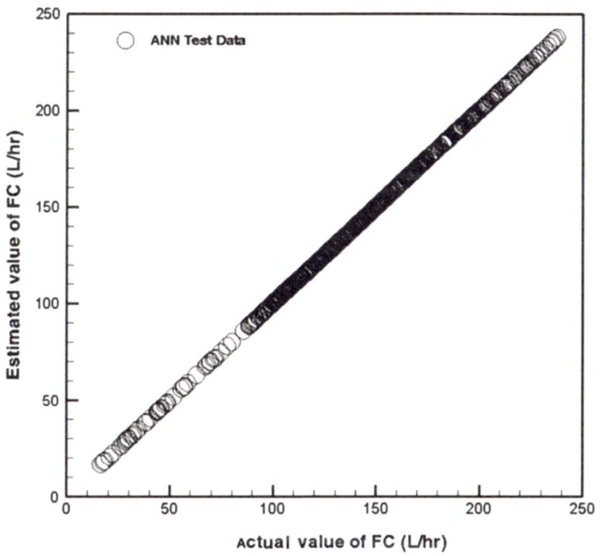

Figure 8.
Comparison of actual values with network outputs for test data (first quarter bisector).

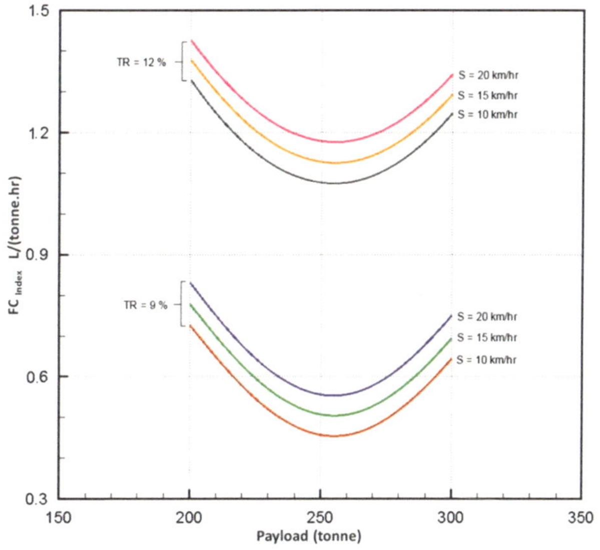

Figure 9.
Correlation between Payload, S, T.R., and FC_{Index} based on the developed ANN model for CAT 793D.

For a standard range of loads, **Figure 9** shows the correlation between Payload, Truck Speed, Total Resistance, and FC_{Index} created by the constructed ANN model for CAT 793D tested in a coal surface mine in central Queensland, Australia.

The results show that ANN could correctly predict the fuel consumed by haul trucks in different conditions. As a result, there are different ranges of consumed fuel for different haul road conditions. **Figure 9** also shows that there is the minimum area for consumed fuel in all tested scenarios. This minimum area is located close to the maximum recommended payload for the truck. It means that loading the truck with the recommended weight can help the mine managers to reduce fuel consumption.

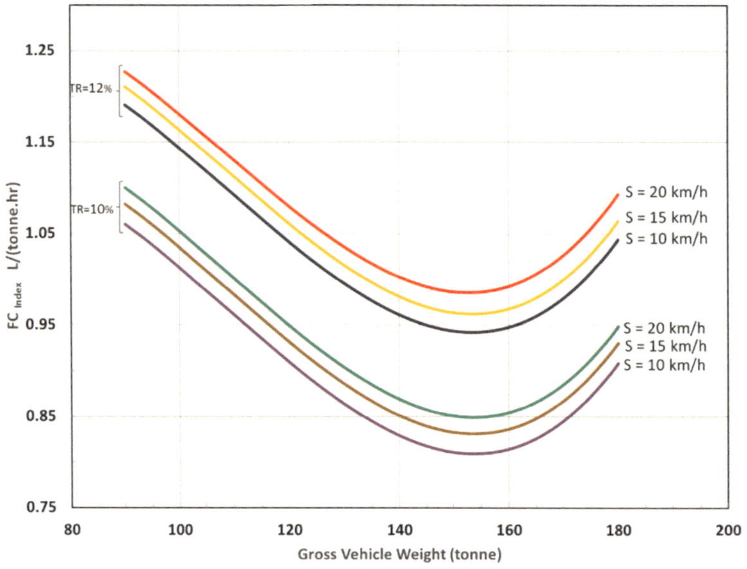

Figure 10.
Correlation between gross vehicle weight, S, T.R., and FC_{Index} *based on the developed ANN model for Komatsu HD785.*

The developed application also tested for a Komatsu truck (HD785) to validate the model for different truck's specifications. **Figure 10** shows the results of model testing for the Komatsu truck.

The minimum areas highlighted by the presented graphs in **Figures 9** and **10** illustrate the potential of deploying optimization algorithms aimed to improve energy efficiency in surface mines. This concept can be a title for further investigations and studies in the future.

7. Conclusion

As old industry mining uses traditional approaches to solve the business problem, energy efficiency improvement is one of the most critical challenges in the mining industry. Mine managers and researchers can benefit from digital transformation and data access by utilizing innovative data-driven solutions such as machine learning and artificial intelligence. This chapter presented a practical framework and developed an artificial neural network algorithm to predict the consumed fuel by haul trucks in surface mines. The successful results of deploying this application in different mines sites have opened a new window for researchers to use the sophisticated AI models to tackle the mining operation challenges. This chapter showed the prediction results for diesel consumption, and it is clear that the prediction is the starting point of advanced analytics. The developed model was used in a large surface mine in Australia and tested on two different trucks (CAT 793D and Komatsu HD785). For both tested trucks, the accuracy of developed mold was reported more than 85%, an acceptable result for a sophisticated AI model that unstructured and noisy datasets have fed. There are more opportunities to use AI to optimize and make decisions to increase energy efficiency in mining engineering.

Author details

Ali Soofastaei[1]* and Milad Fouladgar[2]

1 AI Program Leader, Vale, Australia

2 Department of Mechanical Engineering, Islamic Azad University, Najafabad Branch, Najafabad, Iran

*Address all correspondence to: ali@soofastaei.net

References

[1] Smil V. Energy Transitions: Global and National Perspectives. USA: ABC-CLIO; 2016

[2] DOE, Energy and environmental profile of the US mining industry. Department of Energy. Washington DC, USA: USA Government; 2002. pp. 63-87

[3] Change IC. Contribution of working group III to the fifth assessment report of the intergovernmental panel on climate change. Mitigation of Climate Change. 2014;**1454**:147

[4] Norgate T, Haque N. Energy and greenhouse gas impacts of mining and mineral processing operations. Journal of Cleaner Production. 2010;**18**(3):266-274

[5] Udemba EN, Alola AA. Asymmetric inference of carbon neutrality and energy transition policy in Australia: The (de) merit of foreign direct investment. Australia: Journal of Cleaner Production. 2022;**9**(5):143-149

[6] Cribb J. Australia's energy future. Australian Engineering. 2002;**1**(2):42-48

[7] Golosinski TS. Mining Education in Australia: A vision for the future. 2000; **2**(3):87-92

[8] Energy Account, Australia. 2021. Available from: https://www.abs.gov.au/statistics/industry/energy/energy-account-australia/latest-release.

[9] Statistics, A.B.o., Energy Account Australia, in 2014-2015. 2017

[10] Doe U. Energy and environmental profile of the US mining industry. US Department of Energy Office of energy efficiency and renewable energy. Energiebilanz der Nuklearindustrie. 2002;**4**(2):112-118

[11] Price L, Wang X, Yun J. The challenge of reducing the energy consumption of the Top-1000 largest industrial enterprises in China. Energy Policy. 2010;**38**(11):6485-6498

[12] Doe, U. Mining Industry Energy Bandwidth Study. 2012

[13] Levesque M, Millar D, Paraszczak J. Energy and mining–the home truths. Journal of Cleaner Production. 2020;**84**: 233-255

[14] Peralta S, Sasmito AP, Kumral M. Reliability effect on energy consumption and greenhouse gas emissions of mining hauling fleet towards sustainable mining. Journal of Sustainable Mining. 2019; **15**(3):85-94

[15] Kuzin E, Bakin V, Dubinkin D. Mining equipment technical condition monitoring. In: E3S Web of Conferences. EDP Sciences; 2020;**14**(2):234-241

[16] Doe U. Mining Industry Energy Bandwidth Study. Washington DC: BCS; 2007

[17] Bockasten K. Standards-ISO 50001 helps manage energy costs in mining. CIM Magazine-Canadian Institute of Mining Metallurgy and Petroleum. 2020; **7**(1):36-42

[18] Tasić V et al. Reducing the costs of electrical energy in the copper mining and smelting complex RTB Bor using a distributed control system. In: The 33rd International Convention MIPRO. USA: IEEE; 2010

[19] Soofastaei A. Data analytics for energy efficiency and gas emission reduction. In: Data Analytics Applied to the Mining Industry. London, UK: CRC Press; 2020; **2**(1):169-192

[20] Soofastaei A, Knights P, Kizil M. Application of advanced data analytics to improve haul trucks energy efficiency in surface mines. Extracting Innovations: Mining, Energy, and Technological Change in the Digital Age. 2018;**2**(5): 163-179

[21] Soofastaei A, Davis J. Advanced-data analytic: A new competitive advantage to increase energy efficiency in surface mines. Australian Resources and Investment. 2016;**1**(1):68-69

[22] Arif S. Measuring Productivity in the Australian Mining Sector. Australian Government, Bureau of Resources and Energy Economics: Canberra; 2019. pp. 10-15

[23] Zheng S, Bloch H. Australia's mining productivity decline: Implications for MFP measurement. Journal of Productivity Analysis. 2020;**41**(2):201-212

[24] Hardman D. Coal-mining productivity in South Africa compared with Australia and the USA. Journal of the Southern African Institute of Mining and Metallurgy. 1995;**96**(7):297-301

[25] Kelly G. Renewable energy strategies in England, Australia and New Zealand. Geoforum. 2007;**38**(2):326-338

[26] EEO Analyses of Diesel use for Mine Haul and Transport Operations. 2019. pp. 2-12

[27] Stoett PJ. Global environmental security, energy resources, and planning: A framework and application. Futures. 1994;**26**(7):741-758

[28] Beatty R. Mining truck operations associated with truck-jig apparatus. AIMM Bull. Proc.;(Australia). 1989;**294**(2):75-82

[29] Kesimal A. Applying the queueing theory approach to determine the most economical number of trucks matching shovel for overburden removal in a coal mine. Mineral Resources Engineering. 1998;7(01):29-38

[30] Burt CN. An optimization approach to materials handling in surface mines. Curtin University. 2008;**6**(3):89-95

[31] Kesimal A, Bascetin A. Replacement study of off-highway trucks in an open-pit coal mine in Turkey. Mineral Resources Engineering. 2020;**9**(02):279-286

[32] Komatsu S. Application of Trucks Vol. 2. Komatsu: Tokyo, Japan; 2021. pp. 256-293

[33] Caterpillar C. Performance Handbook. Vol. 2. New York City, USA: US Caterpillar Company; 2021. pp. 184-192

[34] Choi Y et al. Multi-criteria evaluation and least-cost path analysis for optimal haulage routing of dump trucks in large scale open-pit mines. International Journal of Geographical Information Science. 2019;**23**(12): 1541-1567

[35] Tannant DB. Regensburg, Guidelines for Mine Haul Road Design. 2019

[36] Lee T-Y. Development, and Validation of Rolling Resistance-Based Haul Road Management. 2020

[37] Masic S et al. Analysis of parameters about truck transport parameters obtained from different sources in black coal mine "Banovici.". Technics Technologies Education Management-TTEM. 2011;**6**(1):191-196

[38] Soofastaei A et al. Simulation of Payload Variance Effects on Truck Bunching to Minimize Energy Consumption and Greenhouse Gas Emissions. 2015

[39] Soofastaei A. Development of an Advanced Data Analytics Model to Improve the Energy Efficiency of Haul Trucks in Surface Mines. 2016

[40] Soofastaei A, Davies J, Antonio E. Development of a Maturity Scale for Mining Performance and Maintenance Data Analytics

[41] Jeremy, D. EEO energy-mass balance. Mining. 2010;**6**(2):21-28

[42] EEO Driving Energy Efficiency in the Mining Sector. 2010. pp. 18-22

[43] Soofastaei A et al. Energy-efficient loading, and hauling operations. In: Energy Efficiency in the Minerals Industry. Switzerland: Springer; 2018. pp. 121-146

[44] Soofastaei A et al. Development of a multi-layer perceptron artificial neural network model to determine to haul trucks energy consumption. International Journal of Mining Science and Technology. 2016;**26**(2):285-293

[45] Burt K, McShane CLK, Fong OT, editors. Monograph 27- Cost Estimation Handbook. Australia: 2nd ed. Australasian Institute of Mining and Metallurgy; 2020

[46] Pull D. Excavation, loading, and material transport. SME Mining Reference Handbook. 2019;**4**(3):215-221

[47] Runge IC. Mining Economics and Strategy. SME; Tucson, Arizona, USA: Society of Mining Engineering; 1998

[48] Kecojevic V, Komljenovic D. Haul truck fuel consumption and CO_2 emission under various engine load conditions. West Virginia; USA: Mining Engineering. 2010;**62**(12):44-48